**WITHDRAWN
UTSA LIBRARIES**

New Perspectives on South-East Europe

Series Editors:

Spyros Economides, Associate Professor in International Relations and European Politics, London School of Economics and Political Science, UK

Kevin Featherstone, Eleftherios Venizelos Professor of Contemporary Greek Studies and Professor of European Politics, London School of Economics and Political Science, UK

Sevket Pamuk, Professor of Economics and Economic History, The Ataturk Institute for Modern Turkish History and Department of Economic, Bogaziçi (Bosphorus) University, Turkey

Series Advisory Board:

Richard Crampton, Emeritus Professor of Eastern European History at St Edmund Hall, University of Oxford

Vladimir Gligorov, Staff Economist specialising in Balkan countries, The Vienna Institute for International Economic Studies, Austria

Jacques Rupnik, Senior Research Fellow, Centre d'études et de recherches internationales of Sciences Po, France

Susan Woodward, Professor, The Graduate Programme in Political Science, The City University of New York, USA

South-East Europe presents a compelling agenda: a region that has challenged European identities, values and interests like no other at formative periods of modern history, and is now undergoing a set of complex transitions. It is a region made up of new and old European Union member states, as well as aspiring ones; early 'democratising' states and new post-communist regimes; states undergoing liberalising economic reforms, partially inspired by external forces, whilst coping with their own embedded nationalisms; and states obliged to respond to new and recurring issues of security, identity, well-being, social integration, faith and secularisation.

This series examines issues of inheritance and adaptation. The disciplinary reach incorporates politics and international relations, modern history, economics, and political economy and sociology. It links the study of South-East Europe across a number of social sciences to European issues of democratisation and economic reform in the post-transition age. It addresses ideas as well as institutions; policies as well as processes. It will include studies of the domestic and foreign policies of single states, relations between states and peoples in the region, and between the region and beyond. The EU is an obvious reference point for current research on South-East Europe, but this series also highlights the importance of South-East Europe in its eastern context; the Caucuses; the Black Sea and the Middle East.

Titles in the series include:

Ayhan Aktar, Niyazi Kizilyürek and Umut Özkirimli (*editors*)
NATIONALISM IN THE TROUBLED TRIANGLE
Cyprus, Greece and Turkey

William Bartlett, Jadranka Bozikov and Bernd Rechel (*editors*)
HEALTH REFORMS IN SOUTH EAST EUROPE

Vesna Bojicic-Dzelilovic, James Ker-Lindsay and Denisa Kostovicova (*editors*)
CIVIL SOCIETY AND TRANSITIONS IN THE WESTERN BALKANS

Kevin Featherstone, Dimitris Papadimitriou, Argyris Mamarelis, Georgios Niarchos
THE LAST OTTOMANS
The Muslim Minority of Greece 1941–1949

Alexis Heraclides
THE GREEK-TURKISH CONFLICT IN THE AEGEAN
Imagined Enemies

Markus Ketola
EUROPEANIZATION AND CIVIL SOCIETY
The Impact of the EU Accession Process on Turkish NGOs

New Perspectives on South-East Europe
Series Standing Order ISBN 978 0–230–23052–1 (hardback) and
ISBN 978 0–230–23053–8 (paperback)

You can receive future titles in this series as they are published by placing a standing order. Please contact your bookseller or, in case of difficulty, write to us at the address below with your name and address, the title of the series and one of the ISBNs quoted above.

Customer Services Department, Macmillan Distribution Ltd, Houndmills, Basingstoke, Hampshire RG21 6XS, England.

Europeanization of the Western Balkans

Environmental Governance in Bosnia-Herzegovina and Serbia

Adam Fagan
School of Politics and International Relations,
Queen Mary University of London, UK

and

Indraneel Sircar
School of Politics and International Relations,
Queen Mary University of London, UK

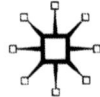

© Adam Fagan and Indraneel Sircar 2015

All rights reserved. No reproduction, copy or transmission of this publication may be made without written permission.

No portion of this publication may be reproduced, copied or transmitted save with written permission or in accordance with the provisions of the Copyright, Designs and Patents Act 1988, or under the terms of any licence permitting limited copying issued by the Copyright Licensing Agency, Saffron House, 6–10 Kirby Street, London EC1N 8TS.

Any person who does any unauthorized act in relation to this publication may be liable to criminal prosecution and civil claims for damages.

The authors have asserted their rights to be identified as the authors of this work in accordance with the Copyright, Designs and Patents Act 1988.

First published 2015 by
PALGRAVE MACMILLAN

Palgrave Macmillan in the UK is an imprint of Macmillan Publishers Limited, registered in England, company number 785998, of Houndmills, Basingstoke, Hampshire RG21 6XS.

Palgrave Macmillan in the US is a division of St Martin's Press LLC, 175 Fifth Avenue, New York, NY 10010.

Palgrave Macmillan is the global academic imprint of the above companies and has companies and representatives throughout the world.

Palgrave® and Macmillan® are registered trademarks in the United States, the United Kingdom, Europe and other countries.

ISBN: 978–0–230–29651–0

This book is printed on paper suitable for recycling and made from fully managed and sustained forest sources. Logging, pulping and manufacturing processes are expected to conform to the environmental regulations of the country of origin.

A catalogue record for this book is available from the British Library.

A catalog record for this book is available from the Library of Congress.

Contents

Acknowledgements	vi
List of Acronyms	vii
Introduction	1
1 Europeanizing Environmental Governance	10
2 The Curious Case of Bosnia-Herzegovina	32
3 Environmental Impact Assessment (EIA) Processes in Bosnia-Herzegovina	63
4 From Pariah to Partner? The Case of Serbia	91
5 Environmental Impact Assessment (EIA) Processes in Serbia	126
Conclusion	150
Notes	163
Bibliography	178
Index	191

Acknowledgements

We wish to acknowledge our deep gratitude to the numerous people in Serbia and Bosnia-Herzegovina, as well as in Brussels, who complied with our request for lengthy interviews. In particular, the help of the following was critical for our research: Florian Bieber, Tanja Bjelanović, Tim Clancy, Miodrag Dakić, Frederic Giovannetti, Miodrag Gluščević, Ljubinka Kaluđerović, Goran Kurtović, Kenan Muftić, Vlado Pandurević, Abida Pehlić, Josip Polić, Vladimir Radojičić, Igor Radović, Ozren Runić, Rijad Tikveša, and Henk Visser.

The research underpinning this book would not have taken place had it not been for the support of the UK Arts and Humanities Research Council (AHRC), who provided generous funding for 'The Environmental Regulation of Foreign Direct Investment in Bosnia-Herzegovina and Serbia: An Assessment of EU and Human Rights Compliance' (ECoBHAS) project (award number: AH/E509673/1). We also benefited greatly from the support provided by the School of Politics and International Relations at Queen Mary University of London.

List of Acronyms

BiH	Bosnia and Herzegovina
CARDS	Community Assistance for Reconstruction, Development and Stabilisation Programme
CEE	Central and Eastern Europe
CEKOR	Centar za Ekologiju i Održivi Razvoj (Center for Ecology and Sustainable Development), Novi Sad
CESD	Centre for Sustainable Development, Sarajevo
CeSID	Centar za Slobodne Izbore i Demokratiju (Center for Free Elections and Democracy)
COM	Council of Ministers
CPV	Centar za puteve Vojvodine (Centre for Roads of Vojvodina), Novi Sad
CSF	Civil Society Facility
CSO	Civil Society Organization
CZZS	Centar za Životnu Sredinu (Centre for Environment), Banja Luka
DEI	Directorate for European Integration
DS	Demokratska Stranka (Democratic Party), Serbia
DSS	Demokratska Stranka Srbije (Democratic Party of Serbia)
EAP	Environmental Action Plan
EBRD	European Bank for Reconstruction and Development
EC	European Commission
ECHR	European Court of Human Rights
EEB	European Environmental Bureau
EIA	Environmental Impact Assessment
EIB	European Investment Bank
EIDHR	European Initiative for Democracy and Human Rights
EMP	Environmental Management Plan
ENGO	Environmental Non-governmental Organization
EU	European Union
EUFOR	EU Force in Bosnia and Herzegovina, Operation ALTHEA
EUPM	EU Police Mission in Bosnia and Herzegovina
EUSR	The EU Special Representative in Bosnia and Herzegovina
FBiH	Federation of Bosnia and Herzegovina
FDI	Foreign Direct Investment
FMOIT	Federal Ministry for Environment and Tourism

GAC	General Affairs Council of the European Union
GIS	Geographic Information System
GTZ	Gesellschaft für Technische Zusammenarbeit (Agency for Technical Co-operation)
HIP	Highway Institute, Belgrade
ICTY	International Criminal Tribunal for the former Yugoslavia
IFI	International Financial Institution
IFOR	Implementation Force, Bosnia and Herzegovina
IMF	International Monetary Fund
IPA	Instrument for Pre-Accession Assistance
IPTF	International Police Task Force
ISPA	Instrument for Structural Policies for Pre-Accession
JNA	Jugoslavenska Narodna Armija (Yugoslav National Army)
LEAP	Local Environmental Action Plan
LLC	Limited Liability Company
MIPD	Multi-annual Indicative Planning Document
MKT	Ministry for Communications and Transport
MOFTER	Ministry of Foreign Trade and Economic Relations
NATO	North Atlantic Treaty Organization
NEPI	New Environmental Policy Instrument
NGO	Non-Governmental Organization
NS	Nova Srbija (New Serbia)
OSCE	Organization for Security and Co-operation in Europe
PADOR	Potential Applicant Dated Online Registration
PCDP	Public Consultation and Disclosure Plan
PE	Public Enterprise
PHARE	Poland and Hungary Assistance for the Restructuring of the Economy
PIC	Peace Implementation Council
REC	Regional Environmental Center
RS	Republika Srpska
SAA	Stabilisation and Association Agreement
SAp	Stabilisation and Association Process
SAPARD	Special Accession Programme for Agriculture and Rural Development
SEA	Strategic Environmental Assessment
SEE	South-East Europe
SFOR	Stabilization Force, Bosnia and Herzegovina
SIDA	Swedish International Development Agency
SNSD	Savez Nezavisnih Socijaldemokrata (Alliance of Independent Social Democrats)

SOG	Savez Opština i Gradova (Association of Municipalities and Cities)
SPS	Socijalistička Partija Srbije (Socialist Party of Serbia)
SRS	Srpska Radikalna Stranka (Serbian Radical Party)
TACSO	Technical Assistance for Civil Society Organisations
UNDP	United Nations Development Programme
UNEP	United Nations Environment Programme
USAID	United States Agency for International Development

Introduction

The audacity and complexity of the EU's mission to transform the successor states of the former Yugoslavia to a point at which they become eligible for membership is well known (Elbasani, 2013; Keil, 2014). Whilst the accession of Croatia in July 2013 was an historic validation of the Commission's Stabilisation and Association strategy, it did little to mitigate the huge challenge that faces those Western Balkan states that remain candidates (Serbia, Macedonia, and Montenegro), or potential candidates (Bosnia-Herzegovina and Kosovo), for whom there is no foreseeable accession date in sight.

Although the difficulties faced by each individual state vary, they all lag significantly behind their Central and East European post-communist neighbours in terms of the political and economic transformation from communist one-party rule and a planned economy to liberal democracy and a market economy. The formation of each of the key stages of democratization has been disrupted: liberal elites have failed to gain a political foothold; nationalist parties have dominated elections and disrupted the emergence of political pluralism and electoral competition (Vachudova, 2003); civil society has been pitched against the state and pilloried by nationalist elites (Fagan, 2010); and privatization, as a driver of state reform and liberalization, has been thwarted by the paucity of foreign direct investment (partly as a result of sanctions in the case of Serbia) and the lingering control of state or quasi-state actors. Those actors and networks committed to reform are invariably too weak to ensure deep-seated change and to dislodge vested interests and those who have everything to lose from substantial reform (Pridham, 2000).

Added to this, they are all weak states with dysfunctional institutions and diminished governing capacities (Kaufman et al., 2010). This is due to the fact that the process of state building – implying the settling of borders, the consolidation of national unity and strengthening institutions – which began with the violent rupture of Yugoslavia, remains incomplete (Elbasani, 2013, p. 13). The successor states remain considerably dependent on the international community for the basic stability and functioning of the state and the economy. Opinion is starkly divided as to whether this continued international presence serves to mediate or perpetuate state weakness (Chandler, 1999). It is certainly the case that the international community's engagement with the successor states is different to its involvement in CEE; the juggernaut of international aid and assistance came to the Western Balkans not to support processes of democratization already underway, but to mediate and end conflict. Donors have remained *in situ* to oversee the transition to democracy and Europeanization. Even in the case of Serbia, where state formation and consolidation is less an issue, the transition from state socialism to liberal democracy has nevertheless been disrupted by nationalism, informal networks (Kostovicova & Bojicic-Dzelilovic, 2008), and a hybridity of private and public sector control of the economy that has limited the transformation of the state's power and regulatory function.

Added to this list of factors that have made the Europeanization of the Western Balkans a far more complex endeavour is a different attitude towards the EU and European integration. These countries, and their citizens and elites, do not identify with Europe in the same way as the Poles and Czechs, for whom a 'return to Europe' was a non-negotiable foundation of their transition from Soviet-style communism. For Bosnians and Serbians, there is a rationale in engaging with EU conditionality, but it is far less rooted in national identity, and the benefits and opportunities are much less obvious. Such relative indifference can, of course, be explained in terms of the lack of substantive progress the governments of these countries have made towards accession, but also the additional conditionality and general lack of enthusiasm for further enlargement amongst member states. In other words, it was not simply that the Poles and Czechs wanted to join the EU *more,* but that their terms of entry were easier and there was widespread support within the EU for taking them in. The Western Balkan states are poorer, less stable, less enthusiastic, and the EU is also considerably less keen on extending further eastwards.

Yet progress is made and a country in the region, Croatia, has entered the EU, and the Stabilisation and Association process (SAp) has thus proved a successful or at least viable route to full membership. Notwithstanding certain differences, there is undoubtedly predictive power in the experience of Croatia for Serbia and even for Bosnia-Herzegovina and Kosovo. Croatia was subject to extra conditionality and compliance with the demands of the ICTY and issues of state building and political reconstruction. Even the most cursory reading of the annual progress reports on Serbia and Bosnia-Herzegovina reveal a *degree* of progress, albeit limited to certain policy or issue areas and occurring at an extremely slow pace. Without wishing to minimize the slow rate of reform and what often appear intangible obstacles to progress, the political scientist interested in understanding Europeanization via enlargement is compelled to address a fundamental question: *When Europeanization does deliver change, why does it happen, how does it happen, and which actors and institutions are critical?* We must also address what is perhaps the more apposite question: *Why does the EU appear to fail to inculcate reform and lock in progressive change in certain policy areas?*

This book is concerned primarily with extending research on what Elbasani refers to as the 'domestic turn' in studies of the Western Balkans (2013, p. 10). The domestic turn refers to a focus on the state and societal level variables within individual countries and across the region that appear to exert a profound impact on success or failure of EU-driven reforms. Whilst the impact of indigenous variables has long been recognized in studies of Europeanization (e.g., Börzel, 2003), studies of the impact of the EU on post-conflict states have tended to focus more on the strategy of the EU and the focus of the international community's 'state building' assistance and intervention. Added to this, much of the early analysis of the region was undertaken by international relations scholars who were keen to use the region as a case study to highlight the perils of externally driven regime change and liberal interventionism. Bosnia-Herzegovina's institutions and political processes were studied through the prism of the Dayton peace agreement and the shared sovereignty model that ensued after the 1992–1995 war. This book is also therefore part of a scholarly shift towards analysing Bosnia-Herzegovina and Serbia from the institutionalist perspective within comparative politics.

This study is interested first and foremost in how change in a particular policy domain (environmental protection) occurs. We start with three important questions:

1. How do newly created institutions develop and engage individuals and networks, and become enmeshed within existing structures?
2. When new institutions successfully challenge or usurp deeply embedded existing structures, how does this happen?
3. To what extent is it the product of EU instruments, or the agency of domestic actors?

Why the environment and environmental governance?

Few now question the extent to which 'post-materialist' concerns for the environment against a 'materialist' backdrop of inflation, unemployment, or post-industrial restructuring is able to capture the political and socio-economic impact of regime change. Indeed, the intrinsic value of a perspective that emphasizes new forms of interaction between state and civil society, and necessitates a fundamental shift in decision-making and regulation has long been appreciated by scholars of Eastern Europe and the former Soviet Union, and too much has now been written that highlights the value of this empirical perspective (Fagan, 1994; Jancar-Webster, 1998; Waller, 2010).

Whereas the early studies of environmental politics in post-communist CEE tended to focus on 'green' civil society and its marginalization during early phases of neo-liberal transformation (Fagan & Jehlicka, 2003), more recent analysis has focused on emergent forms of governance in the new democracies and the impact of Europeanization in particular (Buzogány & Börzel, 2009). The importance attached to environmental regulation by foreign investors and aid agencies (mainly, though not exclusively, the EU) has served to maintain considerable pressure on post-communist politicians to enact reforms, engage non-state actors, and ensure effective implementation. Although progress has not necessarily been steady and sustained in each post-communist state, environmental regulation and new modes of governance have become mainstream political priorities and a fundamental component of post-communist political transition.

When used as an optic to study the Western Balkans and the successor states of the former Yugoslavia, the environmental perspective is particularly revelatory. Beyond what we already know about the difficulties faced by post-communist states trying to enact a battery of complex legislation and implement expensive processes (Carmin & Vandeveer, 2004; Buzogány & Börzel, 2009), it captures the specific complexities associated with a post-conflict transitional setting and the politics of international intervention, where sovereignty is contested and shared. We gain particular insight into the provision of development assistance by international donors and the power dynamic between exogenous and indigenous agendas. We are also offered a lens on the complexities of transforming modes of governance so as to engage state officials and civil society where both lack capacity and are being cajoled into co-operation under the auspices of external pressure (primarily through EU conditionality). We also gain a particular insight into the gulf between communities and the NGO sector, the latter intent on speaking for civil society without necessarily having direct knowledge or experience of popular concerns. In the case of Bosnia-Herzegovina, our research shows how the perspective is illustrative of the problems associated with trying to develop governance co-ordination in a young, fledgling, and divided state in which the power and authority of 'responsible' elites remain contested; where the international community's representatives and domestic elites compete for power and influence, and there is a dislocation between formal responsibility, bureaucratic capacities, and actual power. Beyond simply observing the impact and efficacy of new laws and instruments, the environmental lens bears witness to how the clash of regulatory styles leads to unintended consequences; how the imposition – through EU conditionality – of different norms and costly practices on weak states with diminished resources actually reinforces existing power relations and policy styles rather than driving progressive change. Indeed, what the EU 'locks in' in its intention to install new modes of governance is a particular focus of this research.

But there is a further rationale for focusing on environmental reform in the Western Balkans as a measure of change and political transformation. Environmental decision-making and regulation is often far more politically contentious and agentic in ethnically divided societies than might be assumed. The routing of a new road or the excavation of land owned by an ethnic elite casts a very different light on environmental impact assessment. For example, in the ethnically divided city of Mostar, the proposed route of the planned bypass road had profound implications not just for each community's water supply but for the value of their property and future prosperity. As we discovered throughout the research, this was not simply about groups of citizens being for or against a proposed development: the stakes were far higher and bound up with the ethno-geography of the region. In politically fraught environments, where citizens harbour deep-seated hostility towards elites, the opportunity to participate in a consultation about the environmental impact of a planned development offers a vital chance to articulate a host of political complaints.

The focus and methodological approach of this book

This research employs a neo-institutionalist approach to study the impact of Europeanization on Serbia and Bosnia-Herzegovina from a particular empirical perspective, namely environmental governance. The independent variable is Europeanization, which is understood here as an asymmetrical and politically driven process whereby EU rules, institutions, and mechanisms impact upon applicant and (potential) candidate countries. In seeking to analyse the impact of Europeanization, we adopt a multivariate approach that identifies a variety of domestic and EU-level determinants of change. These include legacies of communism, conflict, and, in the case of Bosnia-Herzegovina, an internationally agreed and imposed constitutional framework that determines the formal distribution of power and authority; the power and influence of domestic elites and networks, as well as what Elbasani refers to as 'deep structures that constrain the capacity of human action to take on and execute EU rules and thus limit the scope of elite-led Europeanization' (2013, p. 6). In tune with much of the recent literature on the transformative power of Europe, we combine rationalist institutionalism (Schimmelfennig & Sedelmeier, 2005b) and constructivist institutionalism (Jacoby, 2004, p. 10; Kubicek, 2003) in an attempt to understand why and how societies and political systems change under pressure from the EU.[1] In other words, we acknowledge that there is both a 'logic of consequences' and a 'logic of appropriateness' at work in explaining change and adaptation (March & Olsen, 1989, 2006).

We use as our empirical focus the legal provision (environmental impact assessment [EIA] laws) in both countries for mandatory public hearings to take place at which politicians, investors, communities, and non-state actors can contest and query the environmental impact of large-scale infrastructure development (such as road building). The perspective provides an opportunity to consider:

formal legal reform and compliance with EU standards (EIA laws);

the capacities of state and non-state actors, and the way they interact;

the dynamics of decision-making, the balance between hierarchy and networks, and the interaction between different state institutions and officials;

the nature of external influence over domestic politics and decision-making; and

how international actors deploy their assistance, tutelage, and influence.

The research, which was funded by the UK Arts and Humanities Research Council (AHRC), was conducted between 2007 and 2011. Although the main empirical focus was the public consultation stage of EIA processes in Serbia and Bosnia-Herzegovina, we were also intent on offering a broader analysis of the capacities of actors, institutions, and processes of governance over a period of time. In order to capture both context and the critical interaction between actors and institutions, we used a process-tracing approach (George & Bennett, 2005; Jervis, 2006; Simon, 1985; Vennesson, 2008), which offers a 'procedure for identifying steps in causal process leading to the outcome of a given dependent variable of a particular case in a particular historical context' (George & Bennett, 2005, p. 176). Process training allows the researcher to determine not just what has happened and how (Jervis, 2006), but also the potential causes and observable outcomes of such interaction and, importantly, the conjunction and the temporal sequence of variables – how they are linked and in what order the impact occurs.

Insofar as process tracing allows for both positivist research – quantifying or illustrating correlation – as well as interpretivist analysis – how and why the link manifests itself – it is particularly appropriate for studying the impact of Europeanization on third countries, where the effects of external pressure are differentiated and become manifest at the domestic level by impacting upon a series of domestic actors, processes, and institutions. In other words, our starting premise is that the involvement of the EU has the potential to transform the interaction between multiple variables, altering the dynamics of power between policy makers and civil society, but this impact will vary according to the particular policy or issue area at stake (level of contentiousness, costs of compliance, and actor capacities), the reform priorities that the EU has identified for that country (what the Commission is particularly concerned about or focusing on), and the degree to which political pressure from the EU is exerted (impetus to ensure compliance and expedite negotiations and accession).

Contents of the book

Chapter 1 offers analysis and a definition of the two key concepts: Europeanization and environmental governance. The objective is to construct a framework for examining the impact of Europeanization on environmental governance in Serbia and Bosnia-Herzegovina that anticipates the type of impact that the EU is likely to exert, based on the experience of Central and Eastern Europe's accession, the experience of South-east European countries that entered in 2007, and Croatia. We aim to capture how the EU understands environmental governance, the combination of hierarchy and networks, what it expects applicant/

candidate countries to achieve, and how. This involves providing the reader with a summary of the tools and mechanisms that the EU employs in the context of (potential) candidate countries, and the various strategies deployed as part of the Stabilisation and Association process (SAp), and the structured dialogue undertaken as part of bilateral negotiations.

The remaining chapters of the book offer detailed empirical analyses of the influence of the EU in the development of multi-level environmental governance. The sections for the two states examined in this book – Bosnia and Serbia – are sub-divided into four chapters (i.e., two for each country). The two countries provide an instructive comparison for the impact of EU strategies in the harmonization of governance. Bosnia-Herzegovina is an ethno-territorially fragmented country with four levels of domestic governance, in which the levels of governance often overlap and conflict with each other. Moreover, prior to the conflicts in the 1990s, it was a peripheral republic, and the wars necessitated an externally imposed process of *state construction*. By contrast, Serbia is a highly centralized state, with Belgrade being the former capital of the Socialist Federal Republic of Yugoslavia. Thus, the reform of environmental governance is predominantly a process of taking existing institutional capacities and resources to channel them into *state transformation*.

For each country, the first chapter provides an historical background and political context of the transformation from authoritarian rule to multi-party democracy. These initial chapters also outline the engagement of the EU in the country, the instruments used by the EU in the process of European integration, and map the various institutional levels for environmental governance. As mentioned above, these structures are more centralized in Belgrade in the case of Serbia, and fragmented across the country in the case of Bosnia-Herzegovina, with weak powers at the centre in Sarajevo. The second chapter in each sub-section provides a fine-grained empirical analysis of the role of key stakeholders during public consultations assessing the environmental impact of major internationally financed, road-building projects. The actors examined in these chapters are representatives from: state-level ministries; sub-state ministries; public enterprises responsible for road-building projects (i.e., roads directorates in Bosnia-Herzegovina and the state-funded company Corridors of Serbia); private and quasi-private technical consultants that completed the EIA studies; environmental NGOs; and local communities.

The final chapter of the book brings together the case studies from the previous chapters to address whether strategies of EU assistance, bolstered by structured tutelage from international financial institutions, have generated the development of new modes of environmental governance *externally*, or whether the variation in the success of this transformation depends more on facilitating and impeding factors *domestically*.

1
Europeanizing Environmental Governance

This first chapter sets out to capture the theoretical and practical challenges of Europeanizing environmental governance in Serbia and Bosnia-Herzegovina. The objective is to construct a conceptual framework for examining the impact of Europeanization on environmental governance that anticipates the type of impact that the EU is likely to exert – based on the experience of the 2004, 2007, and 2013 Eastern enlargements – and to analyze the process of transformation, the agency of domestic and international actors, the role of institutions, and the various constraints that are likely to determine the pace of change.

We need to capture how the EU understands environmental governance, the combination of hierarchy and networks, what it expects applicant/candidate countries to achieve, and how it aims to do so. This involves providing the reader with a summary of the tools and mechanisms that the EU employs in the context of (potential) candidate countries, and the various strategies deployed as part of the Stabilisation and Association process (SAp), and the structured dialogue undertaken as part of bilateral negotiations.

The chapter is divided into two parts. The first offers analysis and a definition of Europeanization as it is understood and operationalized in the context of enlargement and accession. The second focuses on the concept of 'governance' and how the EU understands and develops it in the context of enlargement, and specifically in relation to environmental policy and protection.

Europeanization via enlargement: a definition

Europeanization and enlargement are distinct but overlapping processes. The former can and does occur without there being a prospect of EU membership, as is the case with European Neighbourhood Policy countries. Whilst enlargement is the mechanism through which Europeanization is deemed to take place for countries with a membership perspective, the two processes are still somewhat distinct: Europeanization is a longer-term social as well as political and economic transformation that will occur before and after enlargement has occurred.

Whether associated with enlargement or not, Europeanization is often defined in rather prosaic terms that conjure a diffusion of norms, changes in behaviour, and an element of practical application. It is usually understood as the impact on domestic politics and public policy of EU laws, decision-making, and process of governance (Radaelli, 2003, p. 30). However, even the most cursory knowledge of the eastern enlargements would suggest that Europeanization via enlargement is a multi-faceted and extensive process, worthy of a definition that is suitably broad to enable scholars: to assess the 'varying impact of European integration on domestic arrangements and structures' (Knill & Lehmkuhl, 1999), and to convey the distinctiveness of this impact from other processes such as policy transfer (Bomberg & Peterson, 2000). Thus, we need a definition that expresses more fully the complexities of the interaction between the EU and third parties: how Europeanization both broadens and enlarges domestic processes, and leads to the transfer of new norms and procedures, but 'can also *reduce* policy spaces (and)...also open up others' (Grabbe, 2006, p. 46). A suitable definition must also capture the asymmetric power dynamics of the relationship between the EU and those countries seeking to join. For applicant states the requirements of the enlargement process (absorption into domestic law of the *acquis*) means that the impact of Europeanization is somewhat paradoxical: on one hand it is synonymous with the narrowing of choices and the limiting of policy options, whilst on the other it leads to the expansion of political opportunities, particularly for non-state and civil society actors as well as for local and regional political elites. In other words, the process involves both an imposition *and* a curtailment of options, as well as extended opportunities for domestic actors.

With such considerations in mind the definition of *Europeanization* offered by Radaelli (2000) – in the context of EU eastward enlargement – seems particularly concise and encompassing:

> Europeanization consists of processes of (a) construction (b) diffusion and (c) institutionalization of formal and informal rules, procedures, policy paradigms, styles, 'ways of doing things' and shared beliefs and norms which are first defined and consolidated in the EU policy

process and then incorporated in the logic of domestic (national and subnational) discourse, identities, political structures and public policies.

Radaelli's definition captures the critical notion that Europeanization involves new rule formation and enforcement based on shared values that are considered to be 'normatively superior' (Hughes, Sasse, & Gordon, 2004). But it also conveys changed interaction between state and non-state actors, new behaviours and means of conducting policy making and enforcement.

Thus, the diffusion of norms, values, and behaviours within a candidate state, or indeed a neighbouring country, may well have occurred prior to progress with regards formal compliance. Conversely, an applicant state may have successfully adopted various tenets of the *acquis* and enshrined within the domestic legal system compliant laws, but the process of Europeanization, defined in terms of political process or the way politics is undertaken, may be judged to be limited. Recognizing such a distinction between legal and institutional adaptation on one hand (formal compliance), and political behaviour and process on the other, is critical. This is particularly so in studies of post-communist accession where formal change is (and has proved to be) far more readily and easily achieved than changes in the style and process of governance.

Radaelli's conceptualization of Europeanization offers a framework for measuring both 'soft' and 'hard' impacts. It provides a means of assessing change in 'ways of doing things' as well as in formal rules and procedures, transformation in 'styles, and shared beliefs and norms' (Grabbe, 2006). The two aspects are integral and indivisible in terms of evaluating progressive change, and this research sets out to measure both the importation of new laws and procedures *and* the degree of diffusion and transformation of political processes. The focus is thus on both 'hard' and 'soft' impact, and the emergence of formal policy change as well as transformation in the behavioural interaction between formal/state and informal/non-state actors.

Europeanization via enlargement: the exceptional case of eastward enlargement

Prior to the fifth enlargement (2004, 2007), analysis or discussion of Europeanization tended to focus on the extent to which member states were responding to the growing demands of integration and the constraints on homogenization and enforcement (Radaelli, 2000). However, the process of Europeanization is now immediately associated with the eastward enlargement to include the post-communist Central and East European (2004) and the South-east European states (2007, 2013). Despite the enormous shift in terms of what the EU has set out to achieve in terms of transforming former-communist states with little or no experience of democracy or capitalism, the actual mechanisms at the EU's disposal have essentially remained unchanged. Whilst the EU, unlike most international organizations, has the potential capacity for hierarchical steering (Börzel, 2009c, p. 10), it nevertheless lacks coercive or enforcement powers. Thus, in order to drive substantive change, the Commission relies heavily on the predicted emergence of new modes of multi-level governance in which new forms of decision-making and new constellations of actors will ensure formal and informal compliance under the shadow of hierarchy cast by supranational institutions and by national core executives (Kohler-Koch, 1999; Börzel, 2010). Inducing candidate or potential candidate countries to comply with EU-established norms and requirements is therefore twofold: (i) it involves a behaviour modification strategy based on a process of extensive positive reinforcement, whereby compliance with preconditions is rewarded with additional resources and ultimately full membership, and (ii) the emergence and increased efficacy of new constellations of non-state actors and locations of deliberative decision-making (cf 'network governance' in Kohler-Koch, 1999).

In the context of the eastern enlargement, this mechanism of 'accession conditionality' was used extensively both for securing progress with regard to *democratic* as well as *acquis* conditionality: the former referring to the normative changes outlined in the Copenhagen criteria of 1993 (democratic governance, rule of law, and minority rights); the latter to the incorporation of *acquis*-compatible laws and procedures. Thus, rather than focusing on existing member states' adaptation to EU rules and processes, or the origins of integration, studies of Europeanization via enlargement in the Western Balkans can, courtesy of the accession of the CEE and SEE states, draw on an extensive literature charting the successes and failures of conditionality and the limitations of external governance (Schimmelfennig & Sedelmeier, 2004).

As noted above, the fifth enlargement involved two simultaneous processes: First, the transfer to CEE states of a model of multi-level governance – a way of *doing* politics – that reflected a decision-making ideal that was, to a greater or lesser extent, practiced across aspects of policy making within the EU and amongst existing member states, and seen as characteristic of EU politics (Eising & Kohler-Koch, 1999). Schimmelfennig and

Sedelmeier refer to this aspect as 'what is exported' (2004, p. 662). What must be emphasized, however, in examining the EU's impact across CEE states is that these post-communist countries were very keen to join the EU, and were willing to shoulder considerable costs and enact extensive rules in order to gain entry as quickly as possible (Hyde-Price, 1996). Such was the unprecedented degree of EU influence over the restructuring of the economy, politics, and society of these new democracies that Schimmelfennig and Sedelmeier viewed the eastern enlargement as a process of 'external governance' (2004, p. 661). Second, the CEE states were asked to absorb a huge tome of new laws and directives, to do so in a short time frame, and as part of a regime of continual reporting and extensive conditionality. In other words, the legacy of CEE and its predictive power for other countries engaged in accession negotiations with the Commission must be understood as representing a very specific set of cases, in which EU mechanisms and strategies worked extremely well in a context of a high degree of political will on both sides.

A more detailed analysis of the EU's approach identifies two different mechanisms for achieving change in policy and process: (i) a voluntaristic (or 'social learning') approach whereby a gradual acceptance of EU (or EU-compliant) rules by political elites in non-member states takes place, and (ii) the 'external incentives' model based on the notion that rapid and effective progress towards Europeanization is driven by *conditionality* (Schimmelfennig & Sedelmeier, 2005c). From the latter perspective, states are rational actors and the speed at which EU rule adoption occurs depends primarily on the rational cost-benefit calculations of the domestic political context by non-member elites (Héritier, 2005). This approach to Europeanization is referred to as rationalist institutionalism and is very much the dominant framework within the literature (Börzel & Risse, 2005). The former approach, referred to as constructivist institutionalism, understands change as occurring through the diffusion of ideas and knowledge transfer as a result of twinning processes and the development of transnational networks. Institutional and actor behaviour are modified as a consequence of new institutional arrangements, additional resources, and through tutelage and exposure to different experiences of policy making and enforcement.

Used in conjunction, both perspectives are critical in terms of understanding how and why third countries comply. According to Manners, the EU has effectively established itself as a 'normative power' (2002), and has done so by using financial and non-financial assistance to essentially 'teach' CEE states how to implement European standards and harmonize their practices to the *acquis* (Sissenich, 2007). However, there is much debate as to whether policy elites and civil servants in the Czech Republic or Romania actually had their belief systems and values transformed, or whether they learned how to comply to the point where it became habitual (Parau, 2010; Szulecka & Szulecki, 2013). What is certain is that the new member states are not outliers in terms of their compliance (Falkner & Treib, 2008; Sedelmeier, 2011) and that the provision of additional resources from the EU for state and non-state actors, formal compliance with the *acquis*, and the formation of transnational and pan-European links all played a role in generating progressive change. Whilst there were certainly 'veto players' and serious hurdles, the promise of membership and the effective use of conditionality worked in terms of securing compliance, if not the empowerment of NGOs (Sudbery, 2010), or the emergence of new modes of governance (Börzel, 2009c; Falkner & Treib, 2008).

If the rapid adoption of EU-compliant rules in the run-up to accession was achieved largely through conditionality and the external incentives model, the CEE states also benefited from a period of social learning that preceded the start of sustained 'hard' conditionality (Schimmelfennig & Sedelmeier, 2005a, p. 245). Indeed, prior to the start of accession negotiations and the completion of the *acquis communautaire*, state and non-state actors in CEE were aided by EU programmes for dialogue and technical assistance and provided with opportunities for social and policy learning (Bomberg, 2007; Grabbe, 2006).

Whilst the EU has concurrently employed a mix of strategies to drive change within applicant states, and demonstrated a degree of pragmatism in terms of its use of incentives, rewards, assistance, and tutelage, it has also modified what it expects applicant states to achieve. For example, in the run-up to Bulgaria and Romania's accession in January 2007, far greater emphasis was placed on building administrative capacities and the ability of states to enforce and comply with the laws they had enacted. There was also a particular focus on the rule of law and judicial capacities (Phinnemore, 2013). The experience of the fifth enlargement also suggests subjective rather than rational decision-making with regard to progress towards membership and the start of formal negotiations. For example, laggard candidates (Slovakia in the late 1990s) were invited to begin *acquis* negotiations before democratic conditionality had been met (Schimmelfennig & Sedelmeier, 2005b).

Explaining variation in the pace of reform and compliance

It is imperative when assessing the impact of Europeanization that we should not assume that all change occurring in applicant states is a consequence of EU influence. As Grabbe notes in the context of CEE states, there was a tendency on both the part of the EU and the candidate country to exaggerate the degree of Europeanization for political purposes. The EU obviously is keen to portray itself as 'the principal driver of most reforms' (Grabbe, 2006); the government of the candidate state is equally concerned to demonstrate the extent of its compliance but also 'to blame the EU for unpopular reforms'. It is important also to remember that the EU's political legacy on the CEE states has been largely an *indirect* rather than *direct* one: the EU did not directly bring about democratization in these states, it merely responded to dynamics and processes already under way. It is also fair to say that the commitment on behalf of the CEE governments to Europeanization was unanimous and largely unwavering, expressed at the time of the democratic revolutions in terms of a 'return to Europe' (Hyde-Price, 1996). As will become evident from this study, the situation is somewhat different in Bosnia-Herzegovina and Serbia.

This is indeed an important consideration for assessing the Europeanization of the post-conflict Western Balkans. Although the EU has undoubtedly become the most significant single donor and agent of external governance across the region, it is not the sole actor. The UN, OSCE, the World Bank, and various other American and European development agencies retain a significant presence in the region, working either separately or in conjunction with the Commission, often fulfilling specific peace-building and state-building roles. These agencies set and impose their own conditionality, and whilst there is invariably coalescence in terms of assistance and reform agendas, it is thus difficult to disaggregate the exact impact of the EU, and even harder to measure Europeanization. Any judgement regarding the transformative power of Europe has thus to be placed in the wider context of multiple agencies seeking to steer and engineer change within the region.

In its dealings with the CEE countries, the EU has attempted to influence the development of governance in applicant states in ways that far exceed its official competencies within existing member countries. The theoretical and comparative literature examining the impact of Europeanization and conditionality on applicant states, which is now quite developed, has focused on the transformation of domestic political opportunity structures and the transformative effect of conditionality on the legal and institutional structures of applicant states (see for example, Dimitrova, 2002; Grabbe, 1999; Vachudova, 2005). A much smaller literature exists that focuses on the empowerment of non-state actors and the impact of Europeanization on civil society (Sudbery, 2010).

From both perspectives it is acknowledged that whilst the absorption of post-authoritarian states with barely any tradition or experience of democracy or capitalism has been an overriding success, the Commission's hierarchical intervention has not always succeeded in transforming the process of decision-making and the interaction between state and non-state actors. Indeed, despite the emphasis placed by the EU on building good governance,[1] empowering civil society networks, and the important role of non-state actors, the Commission's enlargement strategy stands accused of having failed to stimulate change, not effectively challenging the incapacities of state as well as non-state actors, and even reinforcing hierarchical and centralized decision-making (Börzel & Buzogány, 2010; Fagan, 2008; Héritier & Lehmkuhl, 2008). For some critics the EU's intervention strategies and the mechanism of conditionality served to actually perpetuate state weakness and prevented new modes of governance occurring by encouraging a 'race to the bottom' in terms of minimal legislative compliance and implementation (Börzel, 2008; Goetz, 2008; Vogel & Kagan, 2004). Conditionality and compliance as the drivers of EU enlargement certainly encouraged applicant CEE states to simply import regulations and processes without necessarily investing in the requisite capacities to enable effective implementation (Buzogány, 2009).

Whilst conceding that the EU has had greater impact in driving formal legal compliance (as opposed to new modes of governance), scholars have sought to highlight and examine incidents of non-compliance and identify additional variables that appear to have determined adherence or non-adherence (Falkner & Treib, 2008; Parau, 2010). Such a 'bottom-up' perspective focusing on domestic constraints has encouraged a multivariable approach to understanding the impact of Europeanization and shifted the academic focus away from a purely rationalist-institutionalist perspective towards one that recognizes the agency of domestic actors, but also the subjectivity of the Commission's approach and the variation across policy and issue areas.

In terms of explaining variation in the pace of change and the degree of compliance, scholars have identified a host of intervening variables. Much depends on the perceived strength of the EU's commitment and the likelihood and imminence of accession (Grabbe, 2006; Pop-Eleches, 2007, p. 142). The costs of accommodating and implementing new laws and procedures, the extent to which the new laws and processes fit with existing practices and traditions, or indeed with reforms already under way, and the opportunities for dissenters and veto players to thwart reform agendas and limit the actions of pro-change coalitions are also critical (Börzel, 2010). The extent to which the momentum for and pace of compliance is heavily dependent on the overall progress of enlargement and accession has also been widely documented from the perspective of particular policy areas and individual states (Buzogány, 2009; Parau, 2009). For the CEE states, significant progress was achieved once a clear timetable and firm date for entry were set.

However, in her study of EU impact on Romania, Parau significantly extends our understanding of the determinants of compliance from the perspective of domestic governmental actors. She employs the term 'compliantness' to refer to the apparent anomalous variation in the willingness of elites to comply across various policy and issue areas. Employing a rationalist perspective, Parau argues that compliantness is contingent upon degrees of uncertainty and the variable impact of competing transnational networks able to influence compliance and non-compliance. What is particularly interesting about this research is that it records instances of compliance where there is little or no pressure from Brussels and cases of non-compliance where conditionality is strong. This immediately confirms the widely held notion that whilst there is contingency between institutional reform and EU conditionality, direct causality cannot be assumed or even directly measured without acknowledging a host of intervening variables. Moreover, a positive correlation cannot be assumed between compliance and the involvement of transnational networks insofar as some networks encourage non-compliance depending on levels of uncertainty and the costs of compliance.

Parau's work forms part of a growing body of literature on Europeanization via enlargement that critiques the more overtly positivist connotations of the early studies, which saw progress towards membership as a rational process of applicant states making measurable advances to clearly articulated and fixed goals. Later studies acknowledge the host of domestic and EU-level variables that interact to determine the pace of change. It is recognized that not only do the EU's goals and criteria change, but the timing of membership is a critical factor which is in turn determined by a host of additional 'internal' factors, such as the capacity of the EU to absorb new members, to undertake an internal restructuring of power, and the political will amongst member states for further enlargement.

Empirical studies of the Europeanization of particular policy areas (e.g., the environment) within Central and East European states tend to offer a more optimistic and nuanced analysis of the impact of conditionality (Andonova, 2003; Bomberg, 2007; Carmin, 2010; Obradovic & Pleines,

2007). We learn that the effectiveness of conditionality is greatest where it triggers the emergence of coalitions of state and non-state actors supportive of change, when it is backed up with clearly focused development assistance, and where the EU is working in tandem with other donors (Bruszt & Vedres, 2009).

However, the perspective of particular policy areas also reveals that there is significant variation in the capacity of applicant states to respond to the host of demands from Brussels; some reforms will be far easier to enact than others, some will be far more appropriate and in tune with domestic political and policy agendas than others, and other reforms will immediately mobilize coalitions of actors opposed to change. But this is not simply a cost-benefit calculation: the 'compliantness' of states and their preparedness to undertake reforms not being purely dictated by the extent to which the EU demands compliance, or the cost of reforms. Change occurs for a host of reasons, some to do with EU pressure, with costs, with historic legacies and existing practices, but also because of the agency of particular networks of individuals. The EU's role is critical in agenda setting and offering a clear depiction of what needs to be achieved in a particular policy area. Beyond that, a host of variables mitigate the Commission's impact.

From governance to environmental governance

Prior to discussing the EU's understanding of governance and, in particular, environmental governance, or indeed how it sets out to transform decision-making and implementation in third countries seeking membership, we need first to establish an understanding of both concepts.

In the sense that a distinction is made by scholars of international relations and public policy between *government* and *governance*, the latter is used here to refer to the notion of multiple locations of power, authority, and control that are both institutionalized and informal, and involve partnerships between state (public) and non-state (private) actors; what Pierre describes as 'the conceptual or theoretical representation of co-ordination of social systems and, for the most part, the role of the state in that process' (2000, p. 3). Emphasizing the primacy of the state is critical insofar as there is a latent (neo-liberal) tendency to assume that *governance* equals the demise of the state rather than a reconfiguration and potentially a maximization of its power and influence (Goetz, 2008).

Inasmuch as governance is borne out of a desire to increase the efficiency and quality of public policy decision-making and implementation (by widening the knowledge base, involving non-state actors and civil society, and private economic actors), it follows that *environmental governance* is generally seen as an attempt to establish greater regulation around matters related to ecological protection through significant interaction between state and non-state actors. At the core of the concept is a recognition that the particular challenge of environmental protection necessitates – perhaps more so than in other policy areas or sectors of human activity – the occurrence of 'a wide set of regulatory processes' including different mechanisms and strategies, but also taking place at different levels – state, sub-state, regional, and international (Lemos & Agrawal, 2006). However, those who employ the concept in an attempt to capture the dynamics of resource regulation and protection usually acknowledge the conceptual and practical tensions between greater devolution of power on one hand, and the internationalization of regulation and control on the other, not least because they are cautious about disregarding the state as a critical and continuing source of regulatory and political power. Most conceptualizations of environmental governance thus identify as critical the co-ordination between levels and the accountability of decision-making regardless of the spatial location of power and authority (Bäckstrand, 2006).

Of course, this has particular resonance in the context of post-conflict states and new democracies, where a particular legacy of transition/ regime change/international administration can be the empowerment of sub-state and supra-state authorities vis-à-vis central government institutions. In such contexts, the normative value of the environmental governance concept lies not just in its ability to legitimate shifting power towards and away from the central institutions of the state, but also to sanction the notion of opening up decision-making and enforcement to multiple actors at *all* levels, as well as emphasizing the importance of fluid interaction and communication.

However, any judgement of the development of environmental governance in Bosnia-Herzegovina (BiH) and Serbia must reflect the realities of existing liberal democracies rather than unrealistic normative benchmarks that will prove unachievable. Shifts towards multi-level governance within established European democracies have not been easy to engender. Indeed, when and where they have occurred, the engagement of non-state actors by the state has been more a response to practical necessity – greater effectiveness and technocratic solutions for reducing risk – than a political or ideological momentum to advance discursive democracy, engage civil society or seek to devolve power away from the central institutions of the state (Sherlock, Kirk, &

Reeves, 2004). Recent critical analysis of the supposed shift from *government* to *governance* in mature Western democracies has revealed that the transformation has, in fact, been exaggerated and that 'there is little, if any, evidence of weakened core executives' (Goetz, 2008). Many of the earlier studies that talked about the hollowing out of the state and the emergence of new forms of power above and below governmental elites (e.g., Rosenau & Czempiel, 1992) stand accused of exaggerating shifts in power and failing to anticipate the resilience of governmental power as a consequence of interaction with private actors and NGOs. Surprisingly, there is also little evidence amongst EU member states of the Europeanization of policy areas resulting in a decline in the powers of national parliaments or governmental elites (Rootes, 2003). In fact, the overriding trend is not, we are told, towards the decline of government and the rise of private actors, agencies, and NGOs; even if such a shift does occur, multi-level interaction is 'monopolised by governments' (Goetz, 2008, p. 4). Despite the normative assertions of the Commission and the emphasis placed on governance within new and applicant states, the policy-making culture within the EU is in fact not one of mixed private-public networks and non-hierarchical decision-making (Börzel & Risse, 2005). The resurgent power of governments and of state capacity within policy making is seemingly overlooked or underplayed by those wishing to export a normative model of multilateral power to acceding and candidate states.

Thus, when attempting to evaluate the impact of EU conditionality on environmental governance in Bosnia-Herzegovina and Serbia, we need to be mindful of what environmental governance looks like across existing member states, and what drives change in behaviours if and when they occur. What we can realistically expect in terms of the scope of change is evidence of a shift towards a *less* hierarchic mode of policy development in which NGOs and non-state actors are more involved and empowered, but not at the expense of state capacity or the ability of the government to cast a 'shadow of hierarchy' (Héritier & Lehmkuhl, 2008). Rather than an expectation of radical multi-level governance, what is sought as evidence of the transformative power of the EU is what Grindle (2004) calls 'good enough governance'; effective government-led decision-making based on interaction between stakeholders.

New modes of governance, the EU, and applicant states

In the context of Europeanization via enlargement, scholars of Europeanization, and the Commission itself, have quite a specific notion of governance in mind. The first point to note is that the

Commission's understanding of governance is essentially functionalist and based on the transfer of rules that 'requires well-functioning institutions of government', as well as 'societal actors ready to engage with the new rules' (Sissenich, 2010, p. 11). Critics would argue that it is a more conventional understanding of decision-making based on institutions and vertical hierarchy as well as diffuse and horizontal networks of non-state actors.

This is indeed reflected in the Commission's attempt – in its quest to drive forth change in third countries – to exert control not just over core executives, but also over judiciaries, bureaucracies, and non-state actors at national and local levels so as to build enforcement capacities and to maximize post-accession compliance (Grabbe, 2001). In this sense Europeanization effectively implies a far-reaching transition from command and control state socialism, to a model of (new) governance based on less hierarchical, multi-level decision-making that is by no means the norm across existing member states (Goetz, 2008).

But how is new governance understood in the context of EU enlargement? Börzel offers a definition of new governance in the context of Europeanization which views the distinctiveness as being about decisions that are collectively binding without being hierarchically imposed, adopted, and complied with voluntarily, and arrived at via processes that 'systematically involve private actors ... (both) for-profit ... and/or not-for-profit' (Börzel, 2009c, p. 8). Although this definition is suitably inclusive whilst at the same time sufficiently measured to mediate between the two extremes of state/hierarchy and non-state/networks, there is little here that immediately appears to be 'new': this is arguably a definition of the realities of public policy decision-making and implementation across Europe as developed over at least the past 30 years. However, what Börzel sees as new is the *extent* to which the model of multi-level governance practiced within the EU and exported as a normative model eastwards stems from (and is inspired by) the peculiar and somewhat unique position of the EU as decision maker and enforcer:

> unlike most international organizations, the EU has the capacity for hierarchical steering. It can adopt policies against the will of individual member states ... unlike a state, however, the EU has no power to enforce its policies but ultimately has to rely on the voluntary compliance of the member states. (Börzel, 2009c, p. 10)

The EU's position thus enables a new model of decision-making and enforcement that relies on multiple actors and voluntary compliance, in theory at least. In essence, therefore, what is new about this mode of governance is the *degree* to which non-state actors and voluntary agreement forms part of decision-making and enforcement; the peculiar institutional and non-institutional arrangements within the EU have essentially inspired a model of engagement that is distinctive and normative.

Externally generated new governance: conceptual overstretch?

How likely is the EU to succeed in exporting this somewhat normative model of governance to acceding and new member states? As noted already, recent research has illustrated the lack of evidence of new modes of governance emerging in the new EU-member states of CEE, and the predominance of conventional hierarchical patterns of decision-making. This has been explained primarily in terms of an absence of *governance capacity* on the part of both state and non-state actors (Buzogány & Börzel, 2010). What has been observed in the acceding states is low levels of co-operation between the sectors that rarely extends beyond consultation, and if it does occur at all, does so only in the context of EU policy (Buzogány, 2009; Guttenbrunner, 2009). Drawing on Mayntz and Scharpf's notion of the 'shadow of hierarchy', Börzel (2007) argues that the critical driver for closer and more effective participation between state and non-state actors is the capacity of policy makers to hierarchically impose outcomes in contradiction, if necessary, to private actors and civil society. In other words, non-hierarchical deliberation and the open exchange of opinion is most likely to occur within a framework in which compliance will ultimately be pursued by state actors or a policy-making elite, who will exercise their authority to impose outcomes. The key to building non-hierarchical modes of governance in which state and non-state actors interact on an equal footing is therefore the fundamental capacity and willingness of elites to ultimately impose reforms and ensure compliance.

This is of immediate relevance to EU efforts to build good governance in candidate and potential candidate countries. Whilst measures to strengthen the capacity of non-state actors are, in principle at least, worthy, governance scholars have argued that the critical driver of new modes of governance is mutual resource dependency: state actors become increasingly dependent on the resources of non-state actors to effect policy change and implementation, and vice versa (Rhodes, 1997; Scharpf, 1978). Where state capacity is severely limited by depleted resources, new modes of governance are particularly crucial insofar as they allows for '(the emergence of) webs of relatively stable and ongoing relationships which mobilise and pool dispersed resources so that collective (or parallel) action can be orchestrated toward the solution of a common policy' (Kenis & Schneider, 1991, p. 36). Moreover, the greater the involvement in decision-making of affected non-state actors, the more likely the implementation and acceptance of policy outcomes, even if the interests of civil society actors have not been fully accommodated or are reflected in the final piece of legislation (Héritier, 2003). Drawing on transaction cost and principal-agent theories, it might be assumed that governments and state actors in post-socialist and post-conflict states, with depleted state capacity, may be more inclined to engage non-state actors because of the additional value (knowledge, expertise) they can bring to decision-making and policy implementation.

However, the flaw in this argument is essentially that when the state is *too* weak, and the capacity of the non-state/NGO sector is also undeveloped, neither side is provided with sufficient incentive to engage the other. Indeed, what is likely to emerge is 'agency capture' or 'problem shifting' whereby authority becomes shared between state and non-state actors with both lacking sufficient capacity (Hellman et al., 2000).

This was less of a concern in CEE states, but poses a significant risk in the Western Balkans. Weak states are unlikely to recapture regulatory capacity once authority and power have been transferred to NGOs or the market. Second, overempowering non-state or private actors, who are likely to bear the costs of compliance, can lead to 'lowest common denominator' policy decisions and solutions, as well as an accountability deficit. Under circumstances in which there is parity in the low level of capacity held by both state and non-state actors, donor-driven attempts to introduce and to stimulate new modes of governance can actually encourage the transfer of authority away from already demoralized and depleted states instead of fostering reform and effective public management. Thus, the discourse of good governance and cross-sectoral partnerships merely obfuscates a neo-liberal reform agenda. Weak states and a sector of weak, or only marginally stronger, NGOs exacerbate the threat of corruption and the persistence of unregulated political and economic spaces.

In terms of engendering a shift towards new modes of governance, it is not simply a question of donor assistance needing to focus instead on building the resources and capacities of both state and non-state actors. Rather, the capacity of the state to enforce policy outcomes is paramount and the outcome of intervention should not be to equate, or level, the capacity of both actors. The state needs to command a 'shadow of hierarchy' over non-hierarchical deliberations if policy effectiveness is to be augmented. If state actors perceive their role to be ephemeral and to be engaging with non-state actors from a position of weakness, then they are unlikely to engage positively in new modes of governance. Equally if there is a sense on behalf of state actors that interaction will result in loss of their agency and autonomy this will equally act as a disincentive for new modes of governance (Hellman et al., 2000). Moreover, if non-state actors distrust the capacity of state actors to be able ultimately to enforce a mutually agreed decision, then this will also act as a disincentive for partnership and governance engagement (Mayntz & Scharpf, 1995).

Whilst the state's capacity to cast a shadow of hierarchy will encourage non-state actors to co-operate and engage in new modes of governance, and the degree to which state actors are capable of resorting to hierarchical modes of governance will increase such willingness, there is a critical tipping point at which the reverse becomes true: the greater the capacity of state actors for hierarchical policy making, the less impetus there will be for co-operation with non-state actors (Mayntz, 2003). This assertion is based on the rationalist claim that state actors seek to maximize or at least maintain their autonomy and capacity within policy processes, and will only relinquish a degree of autonomy to non-state actors if this delivers additional capacity or benefit. In other words, for state actors, there has to be additional value derived from new modes of governance compared to the value obtained from hierarchical and more conventional forms of governance (Kohler-Koch, 1996). In sum, as Börzel et al. note,

> state actors have to possess sufficient capacities in terms of both resources and autonomy in order to cast a credible shadow of hierarchy so that non-state actors have an incentive to cooperate, and state actors are not afraid of being captured. But...these state actors must not be too strong in order to provide an incentive for state actors to seek cooperation with non-state actors. (2008, p. 6)

In terms of understanding why new modes of governance fail to occur or flourish, scholars such as Migdal (1988), Weiss (1998), and Evans (1995) have argued that resources and autonomy alone are not sufficient variables for explaining why change does or does not take place. It is argued instead that much will depend on the specific political culture and dominant attitudes towards state-societal interaction. As Börzel (2003) notes, 'this is particularly true for countries with an authoritarian legacy and no sustained tradition of institutionalized state-society relations'. In post-socialist states the emergence of new modes of governance is also constrained by the fact that non-governmental organizations and other non-state actors are perceived by both the elite hierarchies and the masses to be unaccountable and therefore non-democratic and illegitimate policy makers and enforcers (Howard, 2003; Sissenich, 2007). In post-conflict situations, or in cases whereby state formation and sovereignty remain in their infancy (Bosnia-Herzegovina and Serbia), the prospects for new modes of governance are even more remote.

Notwithstanding the extent to which the EU has or has not been successful in its endeavour, what mechanisms has the Commission employed to stimulate and build new governance? At a basic level, the approach for CEE was based on extensive support for non-state actors. Although the EU's accession strategy for the CEE states primarily targeted central state actors, the Commission's approach for achieving its multi-level governance objectives combined legally enshrining the participation of non-state actors on one hand, with support and assistance channelled through civil society organizations (CSOs) on the other. For many commentators, this has been one of the great success stories of eastward enlargement: transforming local civil society from a disparate array of weak radicalized movement organizations in the early 1990s to professional governance partners capable of shaping and enforcing complex policy frameworks and laws. It is certainly true that various new legal and procedural changes arising from conditionality have served to institutionalize civil society participation. For example, it became mandatory for business and labour interests to be involved in the preparation of new policies relating to various chapters of the *acquis*; pre-accession assistance for applicant states was increasingly subject to government departments establishing partnerships with the third sector and business. Although CSOs from CEE could only participate formally in EU institutions after accession, though their networks and twinning arrangements they could and did alert the Commission to serious breaches of implementation. They were also able to mobilize network resources so as to organize campaigns at the domestic level. In addition to institutionalizing social dialogue across the post-communist states and formally involving non-state actors in policy deliberation, the EU's insistence on informational rights for citizens and civil society actors represented a significant juncture in establishing the foundations for new governance.

In its provision of assistance and tutelage, the Commission's strategy underwent significant change after the late 1990s in terms of a far greater focus on using assistance to achieve compliance, and to use funding frameworks in order to instrumentalize the governance role of civil society organizations as non-state actors. The Commission essentially deployed resources in order to build capacity and thereby engineer participation. Civil society involvement is cast in instrumental terms – to provide new knowledge and ensure as rapid policy development and implementation as possible. From 1992 onwards, the Commission designated a small amount of pre-accession funding directly for NGOs and CSOs. Local delegations of the Commission organized training seminars on how to prepare applications and access funds, and various twinning programmes were launched. Local CSOs were also encouraged to participate in various EU-wide networks, which enabled the transfer of know-how as well as the opportunity to participate in various advocacy forums.

The emphasis placed on capacity building for civil society was certainly not unfounded. During the early years of post-communism, membership of civil society organizations declined rapidly and the fledgling associations that remained active became increasingly marginalized from both politics and policy making. A large part of the problem was that the new generation of self-styled NGOs, despite being in receipt of foreign donor grants, lacked the knowledge and expertise to participate in governance, even on the few occasions when they were invited round the table and asked to comment on policy proposals. Notwithstanding the unwillingness of governmental elites to engage civil society at this time, the NGOs sectors were weak and disparate (Fagan, 2006). NGOs only began to play a significant governance role after 2000, once the accession process had gained momentum (VanDeveer & Carmin, 2004). Thus, in terms of stimulating the governance or advocacy (as opposed to the participatory) function of civil society, the Commission's strategy made perfect sense.

For critics, the ability of the EU to instigate new modes of governance has been much vaunted (Buzogány, 2010; Kutter & Trappmann, 2010). This is in large part due to the fact that the Commission adopts a technical/functionalist approach that equates power with capacities and defines civil society and non-state actors in terms of elite-level professional interest groups able to provide policy expertise. Such organizations congregate around elite actors and the formal decision-making process, rather than channel societal interests or shift the location of power and authority towards horizontal networks and community-based actors. Indeed, the enlargement process arguably curtailed deliberation and public debate (Raik, 2004). Despite the rhetoric of multi-level 'new' governance and multiple-actor involvement, the accession process involves a strong centralizing tendency that reinforces hierarchy, is both paternalistic and technocratic, and is generally more reminiscent of 'old' governance (Kutter & Trappmann, 2010, p. 54; Sbragia, 2000). Insofar as the EU presents its rules and procedures as normatively superior, and the interaction between the Commission and applicant states is highly asymmetrical, opportunities for dialogue with local civil society and for deliberation between state and non-state actors are limited.

As already noted, the EU's external governance approach in CEE stands accused of reinforcing core executives and, somewhat ironically, pandering to the command and control culture of these post-communist states (Goetz & Wollmann, 2001). Civil society organizations in CEE were engaged in a functionalist way, their activities channelled towards policy development and enforcement, and diverted away from deliberative politics. Through expediency and a focus on achieving prescribed outputs, the Commission's approach exported the democratic deficit that has long plagued EU decision-making. Regardless of the policy sector, the key actors involved were the chief negotiators and their small teams of advisers, ministers, and civil servants, who negotiated directly with Brussels. For critics across the region, the process was deemed non-transparent and alarmingly immune from wider political scrutiny. For advocates, the sheer volume of legislation and the rapidity of compliance necessitated fast-track procedures and a leading role for small constellations of experts. It is hard to dispute the assertion that legislatures were undermined and processes of democratic accountability truncated or even completely sidestepped. In large part this can be explained by the overall support in these countries for accession and the desire to move quickly towards full membership. The rapid top-down transposition of new laws and procedures reinforced centralization and the hegemony of narrow policy elites, missing a vital opportunity to establish multi-level governance as a new modus operandi (Grabbe, 2001, p. 1017).

Whether located in the ministry of foreign affairs, or established as a special body embedded within but slightly separate from the higher echelons of the executive, the negotiating team had unique access to information and knowledge. Not surprisingly, therefore, a key concern became the extent to which new knowledge and expertise was being disseminated and whether an emergent epistemic community was likely to absorb a wider array of experts, officials, and politicians. The concern is that instead of EU know-how and expertise being absorbed across government and public administration, what will emerge are 'islands of excellence' in which 'staff working on EU issues have better training and technical expertise' (Grabbe, 2001, p. 1018). As the fifth enlargement progressed the Commission became increasingly concerned about such centralizing tendencies and made deliberate attempts to engage other external actors such as the World Bank and the OECD to help distil new knowledge and co-ordinate the transformation of public administration at lower levels of public administration.

Environmental governance and EU enlargement: what is the EU's record and legacy?

This last section of the chapter focuses on what the EU has achieved in terms of transforming environmental governance in post-communist states. This is offered as a way of drawing together the preceding discussions about Europeanization and new governance, and it is an attempt to offer a clear statement as to what we can realistically expect and anticipate in Bosnia-Herzegovina and Serbia prior to the empirical analysis that follows. This chapter has included an extensive discussion as to the shortcomings of Europeanization and the limits of the Commission's attempts to instigate new modes of governance, but the transformation of environmental policy making and implementation in the CEE states during (if not necessarily since) enlargement has been significant, the scale of the change largely uncontested, and the agency of the EU undisputed.

The development of robust environmental regulation within member states as well as acceding countries is seen as one of the success stories of European integration and a hallmark of Europeanization (Majone, 1996). The environmental chapters of the *acquis*, which amount to over 450 pieces of legislation, constitute an extensive body of directives encompassing recommendations, guidelines, and stipulations (Börzel, 2009b; Jordan, 1999). Aside from the obvious 'green' dividend in community-wide regulation and control, and the economic implications in terms of the single market, the environmental chapters of the *acquis* are also credited with having triggered the emergence of new modes of environmental governance and interaction between elites and civil society within member states (Rootes, 2003). For applicant states, although environmental compliance is costly – particularly at the implementation stage – it is one of the most likely sites for the emergence of new modes of governance interaction (Börzel, 2009b). This is due in part to the relatively high levels of civil society activity concerning environmental protection in post-socialist states (Carmin, 2010), but also because it represents 'an area of positive, market-correcting integration', and therefore the costs involved are deemed to be worthy albeit significant enough to create incentives for co-operation between state actors to engage with NGOs (Börzel, 2009b).

From the perspective of state reform and capacity development, complying with the terms of the environmental chapter of the *acquis* strikes at the heart of the state-bureaucratic reform agenda for post-communist countries: applicant states need to establish new modes of governance and decision-making that are based on deliberative partnerships between state and non-state actors.[2] For non-state actors (and NGOs in particular), the challenges include not just the acquisition of technical/scientific knowledge, but also *transactional* and *participatory* activism capacities (Petrova & Tarrow, 2007).

The CEE states that entered the EU in 2004 (Czech Republic, Poland, Hungary, and Slovakia), and the countries that entered in 2007 (Bulgaria and Romania) faced a number of significant difficulties in appropriating EU environmental standards and adapting their bureaucratic structures in order to implement the new directives effectively (Carmin & VanDeveer, 2004; Parau, 2009). The fundamental problem was not necessarily the absence of administrative capacity or knowledge, but the extent to which limited state capacities and resources could be deployed towards the environment at a time of managing economic transition (Buzogány, 2009). Despite the pre-existence in many cases of quite extensive environmental legislation and trained personnel, co-ordinating and restructuring extended bureaucracies in order to effectively mobilize capacities towards meeting the terms of the *acquis* seems to be the common problem faced by all post-communist candidate states.

What must also be taken into consideration is that whilst Europeanization has certainly exerted a positive impact on the development of environmental governance in CEE states over the long term, many of the critical improvements occurred only in the run-up to accession, when significant pressure was being exerted (Buzogány, 2009). Typically, new laws had been enacted in the early 1990s, drawing in many cases on existing socialist-era frameworks, with little then happening in terms of additional bureaucratic capacities and implementation until the early years of the 2000s (Jehlicka & Tickle, 2004). In other words, there was not, as often implied, a sustained improvement throughout the period under the aegis of progress towards EU membership. What was seemingly critical was the involvement of environmental non-state actors from CEE within trans-European green networks, which provided tutelage for local environmental NGOs in New Environmental Policy Instruments (NEPIs), as well as transferring expertise on lobbying and mobilization (Bomberg, 2007). However, the focus of these large,

Brussels-based organizations has moved away from the Europeanization of future members of the EU towards lobbying EU institutions to adopt certain environmental legislation. The networking of CEE organizations also occurred in part because the local ENGOs from the CEE states had sufficient capacity to actually participate and join such networks.[3]

In sum, the influence of the EU in the realm of environmental protection and regulation in the CEE states has been extensive. The greatest impact occurred in the early 1990s when these states were keen to approximate EU directives and align themselves with environmental practice in member states. This resulted in the transposition of environmental norms into new domestic laws and the introduction of new processes (such as environmental impact assessment). There was often little overt pressure being put on these states by the EU at this stage, other than the promise of membership and the commencement of formal negotiations. The emergence of new actor configurations and the growth of pro-change networks involving NGOs as well as government actors occurred much later (early 2000s) and proved critical in terms of new modes of policy making and implementation that involved NGOs; the EU did much to initiate and nurture this, not least through the provision of assistance to 'green' civil society organizations and the development of pan-European green networks (Bomberg, 2007). There were numerous intervening and contingent variables at work, but the EU accession process set a clear agenda for progressive change and provided resources to allow new constellations of actors and knowledge to emerge. The reality in the new member states today is a mix of policy making and enforcement patterns that combine hierarchical and command and control practices with new forms of decision-making and regulation involving NGOs. This may not constitute 'new modes of governance', but it is not significantly different from how environmental policy and regulation is conducted elsewhere in Europe. This is perhaps the realistic benchmark against which the impact of the EU in Bosnia-Herzegovina and Serbia should be judged.

2
The Curious Case of Bosnia-Herzegovina

Introduction

This chapter will start by providing background information for the complex, post-conflict, and ethno-territorially fragmented country of BiH, which will be followed by a detailed mapping of state and nonstate capacities in multi-level environmental governance. Unlike the other republics within the federal Yugoslavia, BiH had more than one constituent national group, and the 1974 constitution recognized that the republic was home to Serbs, Croats, and Muslims (Bosniaks). After a referendum in February 1992 (boycotted by Serbs) passed to secede from Yugoslavia, the Yugoslav National Army (JNA) intervened in BiH, and hostilities escalated into full-blown inter-ethnic conflict.

The European Community, the forerunner of the EU, recognized the events that were taking place on their doorstep, and had unsuccessfully attempted to mediate in the dispute between the warring sides. Ultimately, the post-conflict Bosnian state was engineered as a result of two internationally brokered peace agreements in 1994 and 1995. The American-mediated Washington Agreement (1994) ended the conflict between Bosniaks and Croats and created the Bosniak-Croat Federation of BiH, which later became the post-war Federation of BiH (FBiH), comprised of ten autonomous cantons. After NATO military intervention, the Dayton-Paris Agreement (1995) ended the war and added the majority Serb entity of Republika Srpska (RS). The Constitution of BiH was included as Annex 4 of the Dayton-Paris Agreement. The details of the constitution were not negotiated by the local protagonists in the conflict, but rather drafted within the US State Department and included as part of the peace agreement that ceased hostilities within BiH.

As a result of this ad hoc process, BiH is a complex, multi-level, power-sharing arrangement between Bosnian Serbs, Croats, and Bosniaks. There are minimal institutional competencies at the central state level. The state-level executive branch is a three-person rotating presidency, consisting of a Serb (elected from RS), and one Bosniak and one Croat (both of whom are elected from FBiH). The main locus of power in the country rests with the two aforementioned state-like entities, FBiH and RS. Both entities have well-developed autonomous political institutions and wide-ranging powers in most policy areas. Brčko District in northeast BiH remained disputed at the end of the war, and it has been ruled as a condominium between the two entities with international administration since 1999 (Geoghegan, 2014). In FBiH, there is an additional layer of ten autonomous cantons at the sub-entity level that resulted from the Washington Agreement. Five of the cantons have Bosniak majorities, three have Croat majorities, and two (Central Bosnia and Herzegovina-Neretva cantons) are considered 'mixed' (i.e., there is no clear majority group). The lowest level of government in the Bosnian constitution is the municipality, with special provisions for the city administrations of Sarajevo and Mostar, the latter of which was imposed by international decree in 2004 to unite the governance in the divided city.

Political progress in post-conflict BiH has been largely steered by the international community in the form of the multilateral Peace Implementation Council (PIC) created after the Dayton-Paris Agreement, which selects and defines the role of the High Representative. The High Representative is the embodiment of the international community in BiH, defined by Annex 10 of the Dayton-Paris Agreement. The High Representative was initially envisioned to oversee and monitor the post-conflict settlement in BiH, but after the 1997 PIC meeting in Bonn, the High Representative was given more extensive powers to unilaterally remove elected officials and to impose legislation. For example, the Bosnian national anthem and flag are both consequences of legislation imposed by the High Representative. Since the High Representative has far-reaching, unchecked powers despite being unelected, many commentators have likened the state of affairs in Bosnia as a neocolonial situation in which external officials take the ownership of Bosnian politics away from local policy makers and citizens. At its worst, the international community's actions in BiH are tantamount to a 'European Raj' (Knaus & Martin, 2003), 'faking democracy' (Chandler, 1999) or creating an international protectorate (Bieber, 2004). On the other hand, there are commentators who admit that the international community has a strong and necessary role in shaping post-conflict BiH (Nowak, 2004); some even argue for stronger and more centralized international community involvement in Bosnian politics (Dizdarević, 2004).

The consolidation of the Bosnian state has gone from a post-conflict transformation shepherded by the international community to a process led by supranational European institutions such as the European Commission. The EU was initially involved in the implementation of certain civilian aspects of the Dayton-Paris Agreement but has had a more central role in the transformation of BiH since 2000. Since the PIC meeting that year, the EU has framed its involvement in and assistance for Bosnia in terms of an entry strategy into the EU through the framework of the Stabilisation and Association process (SAp). EU conditionality for BiH does not just focus on building functioning institutions and economic reforms but also full co-operation with the International Criminal Tribunal for Yugoslavia (ICTY), specific institutional reforms, sustainable return of displaced persons, and human rights. Moreover, the EU has taken on a normative role in BiH by encouraging democratization and regional co-operation through economic and political tools, instead of military ones (Juncos, 2005). Thus, since 2000, the focus has moved away from the goal of full implementation of the Dayton-Paris Agreement towards more Europe-oriented aspirations, changing the engine for reform from 'Dayton' to 'Europe' (Chandler, 2005).

EU instruments used in BiH to build environmental governance

The local institutionalization of supra-state European institutions in BiH commenced soon after the end of the conflict. The Delegation of the European Commission (now the Delegation of the EU) to BiH opened in July 1996 and concentrates on the explanation and implementation of EU policy in BiH, and promoting areas of common EU policy. Moreover, the Delegation administers financial assistance for EU programmes. In addition to the Delegation office, the EU co-ordinates two other institutions in BiH: the EU Force in BiH, Operation ALTHEA (EUFOR); and the EU Special Representative in BiH (EUSR). The Delegation also co-ordinated the EU Police Mission in BiH (EUPM) until its mandate ended in June 2012. EU assistance in BiH to build environmental governance has primarily been deployed through financial and technical assistance, and conditionality.

Financial and technical assistance

The primary source of EU financial help in the years between 2001 and 2006 was the Community Assistance for Reconstruction, Development and Stabilisation (CARDS) programme, which totaled approximately €413 million (European Commission, 2008). The CARDS money supported democratic consolidation through the return of refugees and internally displaced persons (which was the focus of Annex 7 of the Dayton-Paris Agreement), as well as encouraging institutional reform and civil society development. Another source of EU financial assistance has been the European Initiative for Democracy and Human Rights (EIDHR) programme, which builds up social cohesion through the promotion of human rights and rule of law within BiH. The programme started in 2001, and various micro and macro projects have been implemented by local and foreign organizations. Since 2007, IPA has replaced CARDS and other financial assistance instruments[1] in BiH and other candidates and potential candidates of the EU. The objectives of IPA are more focused on shaping the political environment towards eventual accession into the EU, and include: transition assistance, institution building, and regional and cross-border co-operation. The 2013 allocation of funding (the total of these two components of IPA) totaled €108.8 million.[2]

Previous project grants for civil society development have not been specifically earmarked for environmental NGOs. Much of the funding provided by the EU up to the present day has been targeted for the development of legislation compliant with EU standards through multi-level co-ordination between various state and non-state stakeholders. However, more recent strategic documents related to the IPA programme indicate the increasing importance of the environmental sector in the building of governance in BiH. The IPA Multi-Annual Indicative Planning Document (MIPD) for 2011–2013 explicitly listed environment and climate change as a sectoral priority, and the planned IPA funding since 2007 was only exceeded by the financial support for Justice and Home Affairs.[3] The focus for the IPA programme was on legal harmonization, building capacities of environmental institutions, seeking investment for environmental infrastructure, increased awareness of environmental protection during policy making, and further co-operation between state and non-state actors.[4]

To address some of the problems with project-driven capacity building through short-term grants, EU programmes now include technical assistance strategies to build in social learning and tutelage. For example, the Technical Assistance for Civil Society Organisations (TACSO) country office in Sarajevo provides information and advice in six areas: EU policies and NGO funding; relations between NGOs and government; building networks; legal issues (e.g., NGO registration); administration; and project management. These objectives are largely pursued through kick-off events and other training seminars.[5] For the Bosnia TACSO country office, there is an accompanying Local Advisory Group (LAG) that provides opinions on TACSO projects, which consists of representatives from 'prominent' local NGOs, the EU Delegation, entity government, and USAID.[6]

Conditionality

In parallel with technical and financial assistance, the EU has issued a series of documents and signed agreements with BiH that required the fulfilment of certain political and economic criteria as a prerequisite for further and more substantial Euro-Atlantic integration. The first significant document was the 'EU Road Map' issued by Chris Patten (as European Commissioner for External Relations) in March 2000 listing the necessary political, economic, democratization, human rights, and rule of law reforms to undertake before a feasibility study for the Stabilisation and Association Agreement (SAA) could commence. In November 2003, the European Commission approved the feasibility study for the SAA and noted limited progress in the items on the 'EU Road Map', which would pave the way for the start of SAA negotiations.

The EU then set out the parameters of the partnership between it and BiH. In particular, the European Partnership document laid out the short- and long-term political and economic practices to be implemented for continued association between supra-state European institutions and BiH (Council of the EU, 2004a). Moreover, access to assistance and association depended on adherence to the political, economic, and democratization Copenhagen criteria. Non-compliance would lead to the threat of (unspecified) sanctions from the EU. The European Partnership document provided a set of reforms pushed by the EU, specifically customized to address the gaps in the Bosnian political and economic contexts. The European Partnership document from 2004 only included minimal reference to the environmental sector, stipulating that BiH ratify its commitments in the area of environment (among other sectors), and to facilitate the harmonization to EU standards of environmental practice through the creation of a state environmental agency. The European Council updated the European Partnership document in January 2006 (Council of the EU, 2006a) and in February 2008 (Council of the EU, 2008a) based on the annual evaluations by the European Commission. The latter document reflected the findings of the 2007 progress report, and contained a more comprehensive list of stipulated reforms in the field of environment (EC, 2007a). Further to the requirements for a state-level agency for environment and ratification of international conventions (e.g., Espoo and Aarhus), the 2008 document also included administrative capacity building (especially at the state level), better co-ordination between governmental institutions, a more robust system of EIA, and the adoption of a statewide law for environmental protection (EC, 2008a).

The lynchpin of the SAp is the bilateral SAA between the members of the EU and BiH (Council of the EU, 2008b).[7] The difference between the SAA and the earlier Partnership Agreements is that the former is an international treaty between the EU and BiH, while the Partnerships were unilateral. Specifically in the area of environmental governance, the SAA emphasized the need to strengthen procedures and capacities for the development of strategic planning and co-ordination between relevant governance actors in order to approximate the EU *acquis* in the environmental sector. The co-operation should focus in areas of: the reduction of pollution; development of sustainable and clean sources of energy; the effective implementation of EIA and SEA; and ratification of the Kyoto Protocol.[8]

The European Commission monitors BiH progress (or lack thereof) to achieve the objectives outlined in the Partnership and SAA documents. These progress reports change according to the ever-evolving successes and challenges on the ground in BiH. The most recent EC progress reports for BiH have evaluated: horizontal legislation and the approximation of international conventions and EU directives; legislation and implementation in areas of air quality, waste management, water quality, nature protection, noise, GMOs, and industrial pollution control; and development of administrative capacity in the environmental sector (which includes the establishment of a state-level environmental agency and environmental law).

The weakness of EU conditionality in BiH

The eventual signing of the aforementioned SAA between the members of the EU and BiH was not a straightforward process. In 2007, when the EU pushed for the police service to change from a canton-based or entity-based force to one with a centralized command structure based at the state level, weaknesses in EU conditionality became evident. The Bosnian parliament had ratified the police reform bill in April 2007, yet the stated goal of a centralized police force co-ordinated at the state level had failed to materialize. There were two reasons for the international community to demand this type of police reform as a precondition for concluding the SAA. The first is that for a country of approximately 4.5 million people, there were 11 police forces in FBiH and additional services in RS and in the Brčko District with a total of approximately 16,000 officers. These policing agencies acted independently and without co-ordination, and thus there was no unified 'security space' within BiH (Muehlmann, 2008). Secondly, beyond the arguments for operational efficiency of abolishing the sub-state police services, the police reform as centralization had strong symbolic content in undoing the ethno-national separation that characterized post-Dayton BiH. However, such a change to the policing arrangements in RS would be an affront to the very essence of the Bosnian Serb-dominated entity: the decentralized institutions (especially in the security sector) provided an insurance policy to underline Bosnian Serb autonomy within the governmental structures (Fagan, 2010). Despite repeated pressure by the EU tying the SAA to agreed police reforms, leaders in RS remained intransigent, and several deadlines set by the High Representative at the time (Miroslav Lajčák) expired without any progress (BIRN, 2007a). The impasse continued until the end of the year, despite a clear ultimatum from European Commissioner for Enlargement Olli Rehn: 'It's rather simple: no police reform, no Stabilisation and Association Agreement' (BIRN, 2007b). Responding to Rehn's demand, the main political parties from RS and FBiH made a provisional deal in early December 2007 to bolster central policing tasks, such as forensic and training services. Further implementation would depend on the direction of constitutional reform (AP, 2007).

Simultaneous to the crisis with police reform, the High Representative decided to impose a streamlined state-level decision-making process by redefining the Council of Ministers quorum necessary to undertake joint decisions. The High Representative stipulated that the Council of Ministers would be operational if a majority of its members is present and if there is at least one representative from each constituent group. However, this amendment was unacceptable to the leadership in RS, which argued that the lower threshold would open the possibility for representatives from RS, and more specifically, Serb members of the Council of Ministers, to be outvoted. The leader of the largest Serb party in BiH, SNSD, called for representatives to boycott the state organs if Lajčák did not reverse his decision (B92, 2007), and on 1 November 2007, Chairman of the Council of Ministers Nikola Špirić, a member of SNSD, resigned from office (BBC, 2007). However, in an eleventh-hour deal before the 1 December deadline by Lajčák, Milorad Dodik (Prime Minister of RS and leader of SNSD) agreed in principle to the constitutional reform (Tanjug & DPA, 2007), allowing the initialling of the SAA by Špirić and Rehn to take place on 5 December 2007 (Beta, 2007).[9]

Although agreements were reached on both police reform and decision-making in the Council of Ministers, the final outcomes were short of the original European Commission formulations. The final shape of the policing agreement between the parties was in the form of an action plan, and not a more substantive accord for the details of police reform (Collantes, 2009). Similarly, the initialing of the SAA by Rehn rewarded the first steps in starting negotiations on constitutional reform, not the conclusion of such a process. Thus, the events at the end of 2007 could indicate a 'climb-down' from the conditionality initially proposed by the EU, and more generally, the weakness of 'soft power' as a mechanism for externally driven compliance (Fagan, 2010).

Nearly two years after the agreement to start negotiations on constitutional reform, there was still no agreed package of reforms among the political parties from the three constituent groups. Carl Bildt (Swedish Foreign Minister), James Steinberg (US Deputy Secretary of State), and Olli Rehn chaired the meeting with the seven largest political parties; a summit was dubbed 'mini-Dayton', after the talks that ended the conflict in 1995, and underlined how important commentators viewed a resolution to the impasse. Despite the perceived gravity of the situation, these October 2009 meetings in Butmir ended without agreement (Beta, 2009). The issue of constitutional reform was further complicated by a ruling by the European Court of Human Rights (ECHR), which found, in a case brought against BiH by Sejdić and Finci (who are Roma and Jewish, respectively), that the constitution and electoral law violated the European Convention on Human Rights by protecting the three constituent peoples, thus excluding other groups from the Presidency or Upper House. The EU has demanded that constitutional reforms are in line with the Sejdić-Finci ruling, but Bosnian leaders have only started making modest steps towards agreement in December 2013 (Gardner, 2013). Thus, conditionality has been able to bring the political parties around the table, but it has not provided an effective tool for affecting significant change in the critical areas of reform pushed by the international community.

The challenges of Europeanizing BiH

The situation of post-1995 BiH started as peace implementation, followed by a period of peace consolidation, eventually giving way to a phase of building effective democratic governance. All of these stages of political development have been shepherded by international actors: a peace plan brokered by the US government; the implementation of the

Dayton-Paris Agreement overseen by the multilateral PIC; and finally the change of focus from the international community to the EU as the main engine for reform in the development of multi-level governance. This process of Europeanization of the environment and other sectors, as in other post-socialist countries, consisted of two strands: legal or formal compliance with EU standards of best practice through conditionality, and capacity building of NGOs through short-term competitive project grants.

Against this backdrop, Croats, Serbs, and Bosniaks in BiH tend to have three distinct optimal visions for the country. Bosniaks, which comprise the largest national group in BiH, broadly adhere to the EU and international objectives of strengthening the central administrative capacities of the state, so that BiH looks more like a 'normal' state. On the other hand, the Serbs tend to pose the greatest obstruction to the centralization of the Bosnian state. Any centralization of ultimate jurisdiction from RS to the state level has been resisted, especially regarding issues of police and security reform. The Croats are numerically the smallest of the three main national groups in BiH, and a minority in FBiH as well. There were attempts during the Bosnian wars and in 2000 (aided heavily by Croatia in both instances) to carve out a third Croatian majority Herceg-Bosna sub-state entity in West Herzegovina. However, with the Europeanization process in the Western Balkans, the cost-benefit calculus has changed for the Croat community in BiH. Croatia became an EU Member in July 2013. Croats living in BiH are eligible for Croatian citizenship and would thus be able to easily relocate to Croatia and avail in the economic opportunities resulting from EU membership.

The story of BiH is thus a story of three parallel societies within Bosnia, which is reflected in the electoral politics within the country. Critics of the Dayton-Paris Agreement posit that the political institutions that emanated from the peace agreement resulted in deeply entrenched ethno-national politics, in which political elites are rewarded for catering to populist nationalism within their own national group (Chandler, 2006; Joseph & Hitchner, 2008). Elections since the end of the war have been contested along intra-ethnic lines (i.e., there is little or no inter-ethnic co-operation, or significant development of cross-community or non-ethnic political parties as in other post-conflict settings).

In a speech in May 2008, former High Representative Miroslav Lajčák noted the nationalist character of electoral politics in BiH: 'A decade and a half after the war, the vast majority of BiH politicians continue to view almost every issue through a nationalist prism' (quoted in Joseph & Hitchner, 2008). Lajčák also noted the other main problem in post-Dayton BiH: 'Two entities [for] three constituent peoples; five presidents, four vice presidents, 13 prime ministers, 14 parliaments, 147 ministers and 700 members of Parliament, all of whom serve a population of just under four million people' (quoted in Joseph & Hitchner, 2008). In other words, for a relatively small country, BiH is mired in a highly fragmented, decentralized, unco-ordinated, and expensive institutional set-up, with occasionally contradictory and conflicting areas of jurisdiction over important policy matters. Moreover, because the central state is relatively weak, it does not have ultimate authority over most policy areas. The functions of policy co-ordination are largely implemented at the RS and FBiH levels, with minimal co-ordination between the two.

In CEE and SEE post-socialist states, the task of the EU was to take existing state structures and to *transform* them into EU-compliant institutions for environmental governance. By contrast, the EU must first push to *undo* the existing fragmented and unco-ordinated ethnically divided state, and then *rebuild* the state such that there is a well co-ordinated system of EU-compliant environmental governance.

The reform of the Bosnian political system is particularly challenging for three reasons. First, there is a lack of political will among Bosnian elites to bear the political cost of moving away from ethnic politics to reform environmental governance in a way that would be compliant with the bilateral agreements with the EU. Elites prefer to focus on ethnic outbidding to remain in power. Secondly, there is a lack of commitment from the EU itself and non-state organizations working on a European level. When CEE countries started the road to Europe in the 1990s, both the EU and members of transnational environmental NGO networks focused extensive financial resources and expertise to rapidly harmonize these post-socialist states to comply with European legal frameworks, but this is largely absent in BiH. Finally, the capacities of environmental NGOs prior to the post-socialist transition are far lower than in CEE and SEE.

Thus, the multi-tiered institutional structure that resulted from Annex 4 of the Dayton-Paris Agreement not only has profound consequences for governmental action and co-ordination within and across the ethno-territorial divide, but also for the role of non-state actors as credible partners for 'new modes of governance' in BiH, where an interplay of network governance and hierarchy by effective state and non-state stakeholders would lead to good, multi-level environmental governance. The situation in BiH is less encouraging compared with post-socialist states from CEE and SEE that navigated the EU accession, and indeed BiH is a laggard compared with its Western Balkan neighbours (with the exception of Kosovo).

The complex institutional structure of BiH means that there are four levels of local government: the central state, entity, cantons (in FBiH only), and municipalities. The Office of the High Representative does not have any specific responsibility for the environmental sector. The post-settlement configuration resulted in weak state-level structures that are, at best, co-ordinators for the development of governance at the entity level. The main instruments of environmental governance and the potential to implement internationally compliant legislation are only found at the level of both RS and FBiH entities. The cantonal Ministries for Environmental Protection in FBiH do have responsibility for smaller projects, but they are mainly subsidiary to the competencies of the entity authorities. There is no explicit role in the area of environmental protection at the municipal level, unlike in Serbia.

In the non-state sphere, the development of environmental NGOs has occurred mainly since the end of the conflicts in 1995. Unlike their counterparts in CEE, environmental organizations in BiH were not well established during the socialist period. Nor did environmental NGOs become a part of the coalition for reform along with other counter-regime movements. As a result, the development of the environmental sector at the non-state level is at an embryonic stage, and environmental NGOs have relatively low levels of administrative, technical, and financial capacities and are almost entirely dependent on foreign donors (Fagan, 2008; Fagan & Sircar, 2010a).

In the sections below, environmental governance in each 'post-Dayton' tier of government in BiH will be summarized, followed by an examination of the role of non-state actors (including environmental NGOs) in multi-level environmental governance.

State-level capacities

At the state level, there are two institutions – the Directorate of European Integration (DEI) and the Ministry for Foreign Trade and Economic Relations (MOFTER) – that have some role in the transformation of environmental governance. The DEI is a permanent standing body within the Council of Ministers, under the supervision of the Chairperson of the Council of Ministers, which is responsible for co-ordination of activities within BiH related to European integration, such as the harmonization of legislation in line with EU directives. The Director of the DEI is appointed by the Council of Ministers by recommendation from the Chairperson. Although the High Representative imposed the Law on Council of Ministers in December 2002, the first Director of the DEI was not appointed until 2005. Although some commentators have critiqued the DEI as the actual executive authority within BiH (Chandler, 2006), the real powers of the DEI are limited to a co-ordination and adviser role. Along with the analysing of compliance of the legislative framework and draft legislation with EU directives in the area of environmental protection, the Department also provides scrutiny over harmonization in the policy areas of agriculture, rural development, food safety, veterinary protection, and fisheries.

There is no centralized state-level Ministry for Environment in BiH, and what authority is there at the state level functions to simply co-ordinate ministries from the two entities to harmonize legislation with EU standards and international conventions. The state-level responsibility for environmental protection rests within MOFTER, in the Department for Environmental Protection, which is in the Sector for Natural Resources, Energy and Environmental Protection. There was little capacity in the state-level body during the research period: it would be ideal to have a Ministry for Environment or Environmental Protection Agency with a staff of 150 for the types of tasks associated with EU harmonization, but the Department only had six staff members. Moreover, the Department had no budget in addition to the funds to pay the salaries of these staff members.[10]

There are two primary gaps in the state-level configuration of BiH that impede progress towards effective environmental management. The first is the lack of a voluntary, consensus-based, coalition-building process in the Bosnian constitution. The ministerial portfolios, and the deputies in each ministry, are determined by a formula to ensure Croat, Bosniak, and Serb representation and are not the most efficacious manner to implement governmental programmes. The result is that there is no 'joined-up government': ministries act independently of each other with little co-operation between different political parties, especially across the ethnic divide. This leads to a lack of inter-sectoral policies at the state level. Environmental governance affects other areas of state-level regulation, such as transport, energy, or any area that is connected to building of infrastructure, but this not reflected in Bosnian policies (EC, 2009a).

Because of the lack of a state-level agency, ministry, or law for environmental protection, the establishment of a Ministry for Environment, Environmental Protection Agency, or both, have been recommended or requested through various EU agreements and EC progress reports since the start of the process of Europeanization of environmental governance. A unified Environmental Agency was included in an initial draft of the Law for Environment. However, representatives from both entities objected, and posited that if it were necessary to have an agency, there should be one for each entity.[11] The draft law initially had 145 articles but was pared down due to the objections of the entities, and there were only 41 articles in a more recent draft.[12] The Environmental Agency would be easier to establish and implement than a Ministry for Environment, since the latter would require constitutional changes, while the agency would not require any amendments to the constitution. However, as mentioned above, talks have been deadlocked between political party representatives from the three constituent groups, with the round of internationally brokered talks in April 2010 failing to result in agreement. The package of proposed reforms included the creation of Ministries for Environment and for Agriculture.[13]

Despite the relatively bleak picture about the state-level capacities for environmental governance in BiH, there has been marked progress in the ratification of EU directives and international conventions in recent years. In 2009, BiH ratified the following international conventions: the Cartagena Protocol (biosafety); the UN Convention on Biological Diversity; the Convention on International Trade in Endangered Species (CITES); the Bern Convention (European wildlife and habitats); and the Aarhus Convention (public participation in environmental matters); and a law on GMOs to approximate the *acquis* (EC, 2009a). Some of the other conventions or directives are at an earlier stage of implementation, such as the Stockholm Convention (persistent organic pollutants) and the Espoo Convention (EIA in a transboundary context). The process for ratification is long, and initially requires gathering information and opinions from the entity governments and relevant ministries. The convention or directive is then discussed by the state-level parliament, and finally, sent to the presidency for ratification.[14] It is then that measures are taken on planning the implementation of the ratified convention or directive. The area of international agreements and co-operation are the responsibility of MOFTER, but there is not enough staff in the ministry to deal with the demands of formulating an implementation plan for all ratified international obligations, so the work is often outsourced to a focal point in an entity ministry, entity public body, university or private company to lead the procedure for drafting an implementation plan. Due to the weakness of state authority, the relevant entity authorities must approve the focal point.[15] For example, some of the focal points are: RS for Climate Change; FBiH for Biodiversity; the head of a private company that started within a faculty of the University of Sarajevo for the Barcelona Convention; a professor at (West) Mostar University for Ramsar; entity authorities for water management conventions; the Head of the Environmental Department (MOFTER) for GEF; and two other staff members in the Department for the Stockholm Convention and Vienna Convention, respectively.[16] Moreover, BiH published its first countrywide *State of the Environment* report in 2013 (EC, 2013a).

The primary problem with efficient reforms leading to the Europeanization of environmental governance in BiH is that the EU's point of contact and the locus of implementation capacities are weakly connected. The EU, other international agencies, and other countries negotiate with state authorities to conclude treaties. For EU conditionality, the agreement to harmonize governance practices is signed between the member states on the one hand and the state-level authorities on the other. When BiH signed the SAA agreement in June 2008, it was done by Nikola Špirić as the state-level Chairman of the Council of Ministers (Tanjug, 2008a). However, although state officials are required to conclude these agreements, they are at the mercy of the entities and cantons, where the instruments and capacities for policy implementation actually exist. In cases where the entity does not want to engage with state officials, there is little that MOFTER can do. For example, there is an oil refinery near Bosanski Brod, which is in RS and near the Croatian border. Local residents on the Croatian side in Slavonksi Brod have complained that there is a noxious smell from the refinery and were concerned about potential harmful effects. Since BiH has ratified the Espoo Convention, it is responsible for transboundary environmental impacts. To this end, MOFTER contacted the relevant authorities in RS, but the queries were not answered. Croatia then contacted MOFTER, but there was no way for MOFTER officials to push RS to change their practices.

Thus, institutions for environmental governance, particularly implementation capacities, are relatively weak at the state level, even though this is an expectation of Bosnia's bilateral relationships with the EU. Implementation capacities, where they exist, largely lie at the entity level, so it is instructive to now turn to environmental governance in FBiH and in RS.

Entity-level capacities

The Federal Ministry for Environment and Tourism (FMOIT) in FBiH and the Ministry for Spatial Planning, Construction, and Ecology in RS are the competent ministries for environmental governance at the entity level. Both Ministries have an extensive remit, including the protection of air, water, and land, issuing policy strategies related to the protection of these natural resources, and the issuing of permits for major infrastructure projects. This includes highway building projects, thermal power plants, hydroelectric power plants above 5 MW, and extraction industries.

FMOIT has five policy departments, of which the first two are connected to environmental governance: environmental protection; environmental permits; tourism and hospitality; project implementation; and legal, financial, and general affairs. Within the environmental sector, there are four departments: strategic planning and environmental documents; biodiversity and landscape diversity; protection of natural resources and eco-tourism; and the protection of air, water, soil, and waste management. The sector for environmental permits within FMOIT has the following internal organizational units: EIA; environmental permits to prevent major accidents; and registers of polluters, information, and training.

During the research period, there were ten members of FMOIT staff working in the EIA department with occasional assistance from other individuals within FMOIT. This is significantly smaller than the staffing capacities necessary to manage the existing responsibilities under the current EIA legal framework. To comply with the requirements of EIA-harmonized legislation, not only should new installations complete the requirements for environmental permits (which may include a lengthy multi-stage EIA process), but existing installations also need to renew their existing permit to ensure adherence to the newer legislation. Over the past few years, FMOIT has been able to process 40–45 environmental permits annually, but there are thousands of installations within FBiH that will need to be re-examined. The smaller installations will be within the remit of the Ministries for Environment at the cantonal level, but this still requires extensive technical and administrative capacities. The head of the department responsible for EIA studies estimated that cantonal and FBiH authorities needed to process 6000–7000 permits over the next few years, which would require at least 20–25 members of staff at FMOIT.[17] The plan had been to complete all of the permits for existing installations by the end of 2008, but due to the slow pace of processing the permits at the entity and canton levels, the deadline had provisionally been extended.[18] In sum, there was a significant gap between the existing capacities and the administrative expectations of the new legal framework for EIA processes. The head of the department responsible for EIA studies at FMOIT summarized the situation as follows: 'the realization of the plans [for compliance] goes more quickly than the realities on the ground'.[19]

The Ministry in RS is composed of four sectors, of which only the first is related to environmental governance: environmental protection; urban and spatial planning; construction; and project co-ordination, development, and European integration. As of the beginning of 2009, there were only five members of staff responsible for all areas of environmental protection in RS. There had previously been six members of staff in the Sector for Environmental Protection, but one person was promoted to the office of Deputy Minister in Environmental Protection.[20] However, this later increased to eight members of staff in the Sector for Environmental Protection. Although there were no quantifiable data, the lack of staffing capacity was even more acute than in FBiH for two reasons. The first was that there were far fewer individuals in the Ministry in RS compared to their FBiH counterparts. At FMOIT, there were approximately ten individuals working *only* on EIA studies, whereas all of the competencies associated with environmental protection in RS were the remit of seven individuals and a deputy minister. Because of the staffing constraints at the Ministry in RS, there was a designated person to receive all requests for environmental permits, and specific aspects are forwarded to the appropriate official in the Sector depending on area of expertise.[21] The other difference is that there is no intermediate level of cantonal governance in RS, as in FBiH, to handle the EIA processes for medium-sized projects. Since the EIA processes were ratified, most of the permits in FBiH have been issued at the cantonal level.

The lack of adequate staffing not only hampers the implementation of environmental policy within the entities, but also the co-ordination of EU-compliant practices across entity boundaries. There is a consultative forum consisting of representatives from the Ministries of Environment/Ecology from the two entities, and MOFTER is invited as a participant. The inter-entity body meets nearly every month to discuss issues of harmonization. However, there is no requirement for communicating changes in the law from one entity to the other, and more importantly, there are no substantive discussions examining whether the laws are compliant and whether there are better solutions. One example is the Law for Biodiversity, which changed significantly in FBiH, but there was little information or communication about the situation in RS.[22]

However, the most serious problem with the implementation of environmental governance is the dearth of sub-laws around environmental legislation. The laws outline the general parameters and responsibilities for the competent authorities for environmental practices, but the details are contained in the by-laws and rulebooks. For example, the EIA law, which will be examined in greater detail in Chapter 3, outlines the steps to be undertaken in the EIA process, including the public consultation, by the competent authority; but the thresholds of the projects for which an EIA study is mandatory, the responsible level of government (entity or canton), and the contents of the EIA study are contained in rulebooks and by-laws. Although some of this legal documentation exists for EIA legislation, it is largely undeveloped in most environmental areas in both entities. There are entity laws for waste management, air pollution, and water management, but the adequate by-laws have not been passed for many years. The problem was summarized by an official at MOFTER:

> [The laws] are not functional ... They do not have all of the sub-laws. In the first public administration review, they found in 2005 that only about 25 sub-laws have been put in force for the existing laws in the entities. That means that laws actually have no power on the ground ... From 2002–2010, they have changed the laws but have not brought in by-laws.[23]

The panacea for these shortcomings at the entity level is not necessarily to centralize environmental governance, and there are examples from other parts of Europe where decentralized authority is deployed to tackle environmental and other issues in which it is beneficial to bring politics closer to local communities. Given the existing confederal structure of BiH, the most pressing problem in BiH is that, though entities have the resources to carry out more of the functions necessary for policy implementation compared with MOFTER, the capacities in RS and FBiH are still not sufficient to autonomously complete the required harmonization process in the environmental sector. For example, the drafting of an implementation plan by RS officials (as focal points) followed the aforementioned ratification of the Kyoto Protocol. These focal points were required to report their activities to the state, FBiH, and Brčko District, but they have not done this. There was not enough capacity in the sub-state administrations alone to carry out their existing responsibilities and provide implementation plans for international conventions. Thus, Europeanization of the legal framework is the responsibility of state officials in DEI and MOFTER, and the main implementation powers are situated at the entity level, yet neither level of government has the capacities to push rapid implementation of effective environmental governance on its own. BiH suffers from a domestic 'double weakness' whereby insufficient capacities at state and entity levels prevent the development of environmental governance. Although it may not fully solve the capacity gap, the current challenges with harmonization can be partially remedied by pooling capacities and co-ordinating more effectively between state and entity levels, but there is strong resistance at the entity level to ceding power to the centre.

As a result, the lack of resources, severely exacerbated by intransigence regarding state-entity co-ordination, has been detrimental to the implementation of the environmental permitting in both entities. Moreover, the passage of laws approximating EU standards has not been followed by appropriate by-laws and inter-entity harmonization to consolidate the implementation of environmental legislation. Similar capacity problems exist at the cantonal and municipal levels of government, as outlined in the next two sub-sections.

Cantonal-level capacities

The most local units of government with specific environmental competencies are the cantons, which only exist in FBiH. The cantonal level of governance in FBiH provides an opportunity for the decentralization of some environmental governance at the sub-state level without the potential problems associated with the extremely low levels of expertise and capacities in municipal government (which will be the focus of the following section). Moreover, the problems with the implementation of environmental policy in FBiH are exacerbated by competences enshrined in IPCC-compliant entity and cantonal laws, leading to unrealistic deadlines, low levels of administrative capacities, and ever-increasing workloads.

The Deputy Minister for Environmental Protection in the Ministry for Spatial Planning and Environmental Protection in Sarajevo Canton explained the problems with cantonal capacities in FBiH. One of the biggest challenges is that thresholds for projects that require cantonal approval are much lower than in most other European countries and ten times lower than the IPPC standard, so that only the smallest projects are exempt from environmental permits. Thus, FBiH or cantonal authorities are responsible for the permitting of most installations.[24] The problem with setting a low threshold also means that more projects will be subject to the environmental permitting process, which will ultimately be more costly for developers and potentially dissuade investment.[25] In other words, problems with the already-low cantonal capacities are compounded by the European and international standards required for legal compliance. Moreover, Sarajevo and other cantons do not have adequate staffing capacities to process all of the permits for new installations, as well as implementing the permitting process for existing structures.

To estimate the level of existing capacities, Sarajevo and Herzegovina-Neretva Cantons were interviewed in February 2009 and April 2009, respectively. The remaining cantons completed a short electronic questionnaire in June 2010, which was then translated from local languages into English.[26] In Sarajevo Canton, there were six members of staff in the Ministry: the Deputy Minister; a lawyer; specialists for air, nature and water; and an individual responsible for environmental permits. There were also two inspectors. As of February 2009, the role of nature specialist was vacant, and there was not enough money to hire a replacement. As of October 2009, the official responsible for environmental permits had been on extended leave, and the remaining individuals at the Ministry were necessary to complete this work in addition to their primary responsibilities.[27] Thus, even with the full complement of six staff members, there are minimal capacities, but due to financial constraints, the situation in the Sarajevo canton was even more dire in terms of staffing.

The low levels of staffing were also evident in the Herzegovina-Neretva Canton – the second-largest canton that contains the City of Mostar. There were a number of important environmental issues within the canton of national and international importance, including the building of hydroelectric plants along the Neretva, water management in Popovo polje (Ravno), protection of Nature Park Hutovo Blato, and the dispute over the alignment of the Corridor 5c highway between Blagaj and Počitelj. Despite this, the staffing is on a similar level as in Sarajevo Canton. There were six members of staff: air, nature, legal, land registry, statistics, and environmental permits. Additionally, there were three inspectors, but crucially, as in Sarajevo, there was only one official responsible for environmental permits.[28]

The Sarajevo and Herzegovina-Neretva Cantons were in slightly better situations, since they have nearby urban centres with public and private institutions with technical expertise to evaluate the environmental impact of development. Moreover, both Sarajevo and Mostar have entity institutions and relevant public bodies, such as the Public Enterprise Water Management of the Adriatic Sea Basin in Mostar. If there are potential problems with canton/entity co-ordination, it is possible for officials from the canton to easily travel to FMOIT for clarification.[29]

All of the remaining cantons had smaller numbers of staff in the relevant section of the respective Ministries with the exception of Tuzla. The city usually scores quite well on measures of environmental governance, since there are numerous active NGOs, relatively high levels of technical capacities because of the local university, and a cantonal environmental fund. Though there were only six members of staff in the cantonal Ministry responsible for environmental protection, there were three members of staff specifically focusing on the processing of environmental permits, which was more than either Sarajevo or Herzegovina-Neretva Cantons. Still, there was a similar level of monitoring capacity as in other cantons – there was only one inspector for the whole canton. Other cantons have less capacity, in part due to the low numbers of staff. For example, Canton 10 has only one member of staff in the Ministry with competence over environmental matters, there were no inspectors, and the Ministry processed only one environmental permit in 2009. Bosnian Podrinje and Zenica-Doboj Cantons also had only one official each to handle all environmental matters, including the processing of environmental permits. This lack of staff was especially acute in Zenica-Doboj Canton, which processed 35 permits in 2009, nearly the same number as in Sarajevo Canton.

Although there were serious problems with the technical and administrative capacities at cantonal level, there was little evidence that international donors have provided either financial or technical assistance for environmental ministries in the period since 2004. The Ministry in Herzegovina-Neretva Canton is involved in a project with WWF, REC, and a number of local NGOs to protect Hutovo Blato Nature Park. In Canton 10, there were two internationally funded projects: a waste management project funded by the EU and a UNDP-GEF project to protect the biodiversity of Livanjsko polje (worth US$1 million). However, none of the other cantons mentioned any funding or technical assistance received from international donors.[30]

On a positive note, the interview and questionnaire data indicated examples of co-ordination between the canton and the entity-level ministry (FMOIT). For example, Herzegovina-Neretva Canton officials provide feedback and opinions for FBiH infrastructure projects in the canton. Other cantonal Ministries interacted with FMOIT through public hearings, advice, seminars, and drafting of legislation. The surveys from the Ministries also revealed a degree of co-operation between cantons, though the respondent from Posavina Canton indicated that there was insufficient interaction between cantons. Certain cantons specified other cantons with which there has been good co-operation: Bosnian Podrinje

Canton mentioned Herzegovina-Neretva Canton, and Una Sana Canton responded that it had particularly strong co-operation with Tuzla Canton concerning environmental issues.

In general, there were some examples of co-operation between the entity and cantonal levels around environmental issues, as well as amongst cantons. However, the intensity of this co-ordination was relatively low. Canton administration was generally weak, and it is difficult for the small numbers of staff in the Ministries to effectively keep up with legal requirements for new infrastructure, whilst also ensuring that existing installations adhere to the new environmental standards. The consequence of the low levels of administrative capacity and rapidly increasing workload is that cantonal Ministries are marginalized during major projects in FBiH. FMOIT is the lead authority, and cantonal Ministries are treated the same as any other stakeholder, despite the fact that they have legally enshrined responsibility and authority. Moreover, there is little evidence that international donors are providing funding or technical assistance to cantonal ministries in the area of environmental protection. Without the adequate capacities, and scant assistance from donors, it will be difficult for the ministries to achieve substantive compliance in the area of environmental governance.

Municipal level

The lowest level of institutional governance in both entities is the municipality. Each entity has an independent association of municipalities and cities (*savez opština i gradova – SOG*). However, the associations in the entities are weak and meet rarely. The associations themselves have little role in co-ordinating activities or fostering co-operation between municipalities. The main problem is the lack of even basic capacities within many of the municipalities. In the FBiH association of municipalities, there is a board for spatial planning, communal activities, and environmental protection, on which there is one municipal representative from each canton.[31] Questionnaires were sent to officials in these municipalities whose area of responsibility is environmental or ecological protection. There were relevant officers in all ten municipalities, even though there is no legal competence in areas related to environment (which are handled by cantonal and entity authorities).

The level of inter-municipal co-ordination was even weaker in RS. As in FBiH, there is an association of municipalities, within which there is a board for urban planning, communal activities, and environmental protection. However, the board rarely met and had little to do with environmental protection. At the time of the research, the board was chaired by an individual responsible for ecological matters in Banja Luka (an architect by training), and also included the following representatives: a lawyer from Banja Luka, and lawyers from Laktaši, Prijedor, and Mrkonjić Grad.[32] Because of the expertise of the members, the main focus of the board was urban planning. For example, at a board meeting during the research period in 2010, the discussions focused on informal settlements and use of GIS.[33] With the exception of the members from Banja Luka, none of the other municipalities on the board had any ecological functions. Thus, the only respondent to the questionnaire from RS was Banja Luka.

Most of the municipalities surveyed had low levels of staff and little or no experience in deliberative processes. However, the only larger town among the representatives on the board, Kakanj, seemed to have a moderate amount of environmental capacity. Kakanj is an industrial town in Central Bosnia with long-standing environmental impacts due to the local coal mine and cement factory. Since the protection of the environment was a critical issue in the municipality, there was a separate Department for Environmental Protection, Monitoring, and Promotion with four members of staff dedicated to issues related to the environment: the head of the department, two clerks, and an expert associate. Moreover, the municipality government recognized the importance of the environment as an inter-sectoral issue – and there was a sub-Department for Environmental Protection within the Department for Entrepreneurship – and the Departments of Urban Planning and Local Housing ensure that their activities were compliant with environmental standards. The municipal administration in Kakanj was more specifically targeted towards environmental protection than even the administrative structures in Sarajevo and in Mostar. In Sarajevo, issues of environmental protection are handled by four persons: an assistant to the mayor who heads the Public Utility Department and three other individuals, including an expert on environmental issues.[34] In Mostar, the Department of Public Utilities and Environmental Protection had 25 members of staff, but there were only two persons dealing directly with environmental protection.[35] Moreover, there was insufficient inter-sectoral co-ordination. The unified City of Mostar was created by international decree, not by consensual decision-making between Croats and Bosniaks. The combination of departments is determined by ethnic composition, not pragmatic public policy. Therefore, the Departments of Construction and Environmental Protection were separate, and there was little co-operation between the two.[36] None of the other municipalities that were surveyed had a department that included environmental issues. However, representatives from both of the city administrations mentioned that most of the competencies for environmental matters were at the canton and entity levels, so the levels of staff were sufficient for the constitutionally defined position of city-level/municipal self-government. Moreover, giving municipalities competence over environmental or other policy matters to include local communities needs to be balanced with the vulnerability to corruption and clientelism at the municipal level versus higher institutional levels of government.[37] In RS, there was no dedicated Department for Environmental Protection in Banja Luka, but there was a section within the Department for Public Utilities with five members of staff working on ecological issues.[38] In Trebinje, there was a single individual (an expert in forestry) who worked on matters related to the environment, including waste management and hygiene as well as contact with NGOs and media on relevant matters.[39]

In some of the smaller municipalities across FBiH, there was insufficient staffing capacity to have a dedicated official for environmental protection. For example, officials for environmental protection were also responsible in their respective municipalities for: tourism and economy in Foča-Ustikolina; reconstruction, fire protection, water resources, and infrastructure in Drvar; and local infrastructure, water management, and forestry in Drvar.

In addition to the capacity for regulating environmental matters, there was also some variation among FBiH municipalities regarding environmental inspection. Both of the city administrations (i.e., Sarajevo and Mostar), as well as Drvar and Kakanj, had inspectors that were independent of the municipal Departments for Environment. However, in the municipalities of Foca-Ustikolina, Bužim, and Čelić there were no current inspectorate services.[40] In Ljubuški, Jablanica, and Donji Vakuf, one or more of the officials responsible for environmental matters were also inspectors. Banja Luka, the largest town in RS, and the only one with a moderate level of capacity, does not have a separate inspectorate. Two of the five individuals responsible for ecological matters were also environmental inspectors.

With the varying levels of capacities, it is unsurprising that the degree of involvement in processes such as EIA public consultations differs amongst the municipal governments. Surprisingly, the relevant officials in Sarajevo and Mostar have not directly been involved in EIA processes. By contrast, for EIA process for the Corridor 5c highway and Block VIII of the local cement factory, representatives from Kakanj municipality were involved in providing non-technical summaries of EIA studies to all relevant stakeholders, informing local stakeholders about the public consultation, making the municipal department officials available to the local community for questions, submitting a written opinion about the EIA, and hiring a technical consultant to provide information about the EIA to municipal government. The municipality of Donji Vakuf provided data on natural freshwater springs, cultural monuments, and other baseline information during the EIA processes for the road between Lašva and Travnik, as well as a nearby mine and the thermal power plant in Bugojno. None of the other respondents indicated any EIA involvement since 2004. The Banja Luka municipality had not been a stakeholder in an EIA public consultation process since 2004.

Some of the municipalities have also drafted and/or adopted Local Environmental Action Plan (LEAP) documents, which requires a dialogue between state and non-state actors, and is an indicator of at least a modest level of multi-level environmental planning and governance. The cities of Mostar and Sarajevo had been involved in such processes, as well as the municipalities of Kakanj (adopted in 2010) and Jablanica (adopted in 2003).[41] The six other FBiH municipalities surveyed had no LEAP at the time of the research. Banja Luka completed the LEAP in 2010, which was funded by the municipality. Thus, though some of the municipalities had undertaken this process, most of them had not demonstrated the minimal capacities to draft and adopt a LEAP.

As in the cantonal governments, the primary problem was a basic lack of capacities, and a lack of targeted international donor assistance, either financial or technical, in the area of environmental protection. The only significant donor initiative for capacity building for governance at the municipal level was the GAP programme funded by USAID, but it only dealt tangentially with municipal capacity to manage environmental policies. Of the FBiH municipalities interviewed or surveyed, Foča-Ustikolina received UNDP funds for equipment in 2009; Donji Vakuf received World Bank funds (2009) for a sewage system, UNDP funds (2008) for cleaning the river, Oboracka, and REC training (2008) to design feasibility studies for the upper Vrbas; Mostar municipal officials attended training seminars by UNDP and REC, and the private engineering firm CETEOR; and the expert associate from the City of Sarajevo participated in workshops organized by REC and SIDA. All of these activities were one-off training or funding and did not contribute to the long-term capacity building for inclusion in environmental governance. In that way, some of the problems at the municipal level mirrored those in the non-state sector, which will be examined below.

Non-state actors

The development of effective environmental NGO capacities as part of multi-level environmental governance in BiH suffers from at least five challenges. **First**, the lack of a politicization of nature protection in the process of regime change in most parts of Yugoslavia during the late 1980s has stymied the creation of a national environmental movement. **Second**, the effect of ethno-national violence and territorial separation among the three constituent peoples shape the political spectrum, not the left/right cleavages predominant in the rest of post-socialist Europe. **Third**, technocratic/scientific interpretations of environmental protection are privileged over framing environmentalism in terms of democratization and the inclusion of civil society, particularly in the context of consultative processes like EIA. **Fourth**, unlike in CEE countries, there have been relatively low levels of foreign direct investment (FDI) and the pace of privatization has been slow. Consequently, there has not been the growth of a post-materialist middle class that could potentially support environmental causes. **Finally**, transformation of environmental governance is inextricably linked to the rigid process of Europeanization, which narrows the set of environmental issues for which NGOs can receive financial support from international donors (since there are few domestic sources to address other subjects).

As a result of these factors, Bosnian ENGOs have very low capacities, even compared with the situation elsewhere in the region. A large proportion of NGOs carrying out work related to environmental protection have no paid staff, no budget, and few members of staff able to communicate in English. The latter is a prerequisite for any European funding, since applications for EU funding, along with the concept note and other supporting documentation, must be in English (Fagan, 2008).

The Regional Environmental Center (REC), though not an environmental NGO nor a major international donor, acts in some ways as both a national environmental NGO and donor in BiH and other post-socialist countries in CEE and SEE. REC was established in the 1990s to oversee and co-ordinate the development of environmental civil society organizations in the region and is headquartered in Szentendre, Hungary. Much of its funding comes from Hungarian and American governmental sources, but it is an important actor in local environmental civil society and has local offices dotted throughout the region. The REC country office has kept a list of Bosnian environmental NGOs, which was created through the inclusion of organizations that had attended REC activities, received the REC newsletter, and through personal contacts.[42] In 2001, when REC conducted the previous survey of environmental organizations, there were 127 Bosnian organizations on the list, of which there were 70 respondents. REC compiled the 2006 directory of environmental NGOs by sending a questionnaire in local languages to the 200 organizations on the original list. There were only 87 respondents, and many of the organizations that responded were either no longer operational or were NGOs predominantly involved in other issues and campaigns with an environmental component.[43]

With offices on the ground in BiH, REC is well placed to make general evaluations about the state of environmental civil society in BiH. The Director of REC (BiH) claimed, when interviewed in 2008, that the EU and other international donors favoured larger organizations at the expense of smaller ones and that there were only 25–30 such larger organizations.[44] The President of Ekotim, an environmental NGO that is able to implement national campaigns and has EU funding, posited that there were no more than five other environmental NGOs within BiH which have the level of capacities to partner with Ekotim and carry out extensive regional or national campaigns.[45] The President of the Centre for Environment (CZZS) also claimed that it was difficult to link with reliable NGO partners with sufficient capacity, and cited problems in finding partners to draft a strategy document for the energy sector in BiH as an example.[46]

The REC questionnaire data reflect two general trends. First, most NGOs had little or no budget. Of the NGOs surveyed, 14 had no annual budget, and 35 organizations attempted to carry out their activities on less than €5000 per year. By contrast, there were seven NGOs with annual budgets of over €100,000. This represented an increase from the number of organizations with such a high annual budget from the 2001 questionnaire. However, there were also a larger number of NGOs in the 2006 survey with little or no budget, indicating a widening gap amongst environmental organizations in terms of financial capacities (REC, 2006). The REC data also indicated that the environmental NGO sector clustered around certain regional centres, particularly Sarajevo (14 NGOs), Banja Luka (9 NGOs), and Tuzla (8 NGOs).

The problems associated with the establishment and consolidation are not dissimilar to movements in CEE. In the countries from the first wave of post-socialist accession to the EU, the process of democratization removed the counter-establishment activism from most civil society organizations, and instead, they became co-opted into the process of transition (Baker, 1999; Lomax, 1997). Because of the demands of international donor funding, the more successful environmental civil society organizations have been those that are available for service provision to aid in the process of Europeanization (Hallstrom, 2004; Hicks, 2004). Thus, the agenda for environmental governance pushed by the EU is based on assistance governmental structures in compliance, and smaller grass-roots initiatives are neglected (Bell, 2004). In other words, if a project does not fit within the parameters of EU harmonization, it will not be supported. In BiH, the environmental NGOs that have been supported by the EU and other foreign donors are those that have advanced administrative and project management capacities, not necessarily organizations with strong links to local communities (Fagan, 2006, 2008).

A further problem with environmental NGOs in BiH is that many of the organizations were not really non-governmental. Individuals linked to public institutions, research institutes, and universities creating NGOs to provide an alternative revenue stream is not unique to BiH and was also the case for donor funding across post-communist Europe and successor states of the former USSR (Mandel, 2002). Those with access to the existing technical resources available at university faculties and governmental institutions have an advantage in applying for and obtaining EU funding. For example, a university professor who is also the director of the Hydro-Engineering Institute of Civil Engineering Faculty, University of Sarajevo, heads the Centre for Sustainable Development (CESD). The CESD mainly focuses on assisting public and private sector partners to meet environmental benchmarks through the use of greener technology (Fagan, 2008). Individuals at the Faculty of Civil Engineering established the CESD as a way of accessing project funding for Bosnian civil society organizations. With the technical and statistical expertise of the university at its disposal, CESD remains one of the most successful organizations in BiH in securing project funding from the EU (Fagan, 2008).

In addition to the challenges faced by ENGOs that are similar to those faced by CEE and SEE organizations, a number of contextual factors – both domestically and internationally – explain the perennial weakness of Bosnian environmental civil society compared with organizations in other parts of post-socialist Europe. **First**, since the transition was from a socialist to an ethnified consociation, environmental issues have been marginalized in favour of ethno-nationalist ones. Environmentalists were not part of a broad pro-European coalition at the point of transition, because this 'transition' was negotiated amongst representatives of the military conflict, unlike any other country in the region. Therefore, it was not possible for environmental issues in BiH to be used as vehicles of transition as they had been in Hungary and Bulgaria, which provided a longer legacy of environmental activism in those places.

Second, and relatedly, this has also meant that environmental civil society organizations did not develop sufficient capacities and were not established long enough to be credible partners on a transnational European level, either during or after the socialist period.[47] Although part of the problem lies with the capacities of Bosnian NGOs, the commitment of pan-European actors can also be contrasted with the situations during the Europeanization of environmental governance in CEE and SEE (Bomberg, 2007). In addition, pan-European NGOs that form a part of the 'Green 10' network of organizations lobbying in Brussels have not been as active as they had been in the integration and accession of the CEE and SEE countries. Members of the Green 10 were pivotal in a tutelage role for local NGOs in CEE to train them about New Environmental Policy Instruments (NEPIs). However, the focus of these large Brussels-based organizations has moved away from the Europeanization of future members of the EU towards lobbying EU institutions to adopt certain environmental legislation. For example, in the years before CEE states joined the EU, Transport and Environment (T&E) had a programme on EU accession. However, this programme has since ended and the organization now focuses on what it 'does best' (i.e., lobbying EU institutions to ratify standards related to transport policy).[48] The EEB also carried out extensive training sessions during the accession of CEE states. The SEE states that joined the EU in 2007 were also a part of these programmes. However, EEB no longer carries out these activities, because local organizations do not have a threshold level of capacity; but also EEB objectives at the time were not focused on EU integration and Europeanization, though there have been members added since then from Macedonia, Croatia, and Serbia.[49] WWF was visible and active in all parts of South-east Europe, including the Western Balkans and is implementing programmes for water management along the Neretva River. Although capacity building is an integral part of the interaction between WWF and its local partners, this is not necessarily tailored for Europeanization.[50]

The **third** issue is that in CEE, the influx of FDI following the post-conflict transition allowed for a new middle-class population to be able to donate money to post-materialist causes such as environmental protection. This created a variety of revenue streams for environmental organizations, and reduced the over-dependence on international donor funding. Ultimately, this empowered ENGOs to have a more visible role in post-transition politics. By contrast, there is far less FDI in BiH, and the economic situation means that fewer people have the means to donate to post-materialist causes. Bosnian ENGOs are thus unable to raise sufficient funds through membership dues and are thus overwhelmingly at the mercy of international donors as their sole means of survival.

The **fourth** and most important reason that BiH is a laggard in environmental civil society development is the ethno-national post-conflict fragmentation within the country. The absence of a national environmental NGO network is partially due to the fact that there is not a national environmental authority with which to interact. As mentioned above, the strongest environmental authority (and governmental authority more generally) is at the entity level, so leading environmental NGOs have opted to create two eco-networks (Eko-Mreza) of civil society organizations, one in FBiH and the other in RS. Since there are already only a few capable organizations in the country as a whole, dividing the network in half further weakens the prospect for such a network to be a significant force in Bosnian politics.

The problems of ENGO co-operation under conditions of ethno-territorial fragmentation can be illustrated in the area of water management. Because of rapid and often unchecked infrastructure development, including the damming of rivers and smaller streams for hydroelectric power, one of the most salient issues for BiH is the proper management of its water resources. The demand for electricity is relatively low per capita within the country, so hydropower plants are springing up so that the owners can export 'green' energy to other countries, even though there are potential negative environmental effects associated with changing the flow, such as excess salinization of the soil. One of the regions in which the ecological balance is quite sensitive is the karst area in Herzegovina, Croatia, and Montenegro. Proper protection of the landscape requires co-operation across international borders as well as inter-ethnic entity borders. Although it is an important issue, NGOs in Trebinje have encountered difficulties co-operating with organizations from outside RS (Fagan & Sircar, 2010b).

Conclusion

As in other parts of post-socialist Europe, formal environmental compliance has led to a 'race to the bottom', where legal standards are adopted just beyond the threshold for EU minimum benchmarks without sufficient capacities for governmental institutions to monitor or implement these standards, and the rapidity of which sidelines the civil society sector and fails to trigger 'new modes of governance'. In the non-state sector, the pressure for rapid harmonization and lack of relevant expertise in governmental institutions has resulted in a situation in which ENGOs have developed mainly project management capacities, and the non-state sector has been co-opted to become service providers for Europeanization, whilst simultaneously being severed from local citizens. These problems are exacerbated in BiH, when comparing Europeanization of environmental governance with so-called successful cases from CEE. First, and most importantly, the reforms that stand in the way of further European integration are not only regulatory/institutional as in CEE countries, but there were post-conflict requirements as well, including minority return and co-operation with the ICTY in the apprehension of war crimes suspects. A second and related obstacle to effective Europeanization is the social and institutional fragmentation resulting from the wars in the 1990s, which are the divisions that have been institutionalized in the complex, ethnified constitutional structures in BiH. Thus, there is a weak centre with little power to co-ordinate or adjudicate between municipalities, cantons, and entities around environmental issues of common concern. The conflicts also left BiH as a patchwork of local mono-ethnic clusters unable and unwilling to cross ethno-national lines to co-ordinate activities; this local fragmentation impedes the development of a national environmental movement. These problems are further compounded by the lack of EU assistance and lack of attention from pan-European NGO networks. Although the EU has invested heavily in BiH, including some aspects of building environmental governance, it is not nearly as high or as sustained as with the post-socialist states that joined the EU in 2004 and 2007. Moreover, other non-state environmental civil society networks that played a social learning role in CEE and SEE are largely absent in BiH, and the priorities of these pan-European NGO networks have largely moved away from fostering relationships with local BiH partners and are focusing instead on lobbying in Brussels. Not surprisingly, the factors outlined above have led to the continued weakness in areas of environmental governance at the state, entity, cantonal, and municipal levels. These factors have also resulted in a small core of moderately effective Bosnian ENGOs aiding in governmental service provision through donor-funded projects, without any evidence of significant development of local NGOs or of a national green movement as a counterweight to the government in building robust multi-level environmental governance.

However, even with these challenges, there are two reasons for some hope for a more positive prognosis for the long term. First, there are examples of local or regional co-operation between state and non-state actors around issues of common concern. For example, the dispute surrounding the alignment of Corridor 5c between Počitelj and Blagaj has triggered a regional cross-ethnic response from NGOs in BiH, Croatia, and Serbia. The case study of the Počitelj-Blagaj section of Corridor 5c will be examined in greater detail in the next chapter.

The second reason for cautious optimism is that there seem to be lessons learnt, especially by the EU, on ways to foster new modes of environmental governance during the process of capacity-building compliance in environmental issues. The EU has started to pursue a more structured 'social learning' approach with study visits for representatives of NGOs, and a locally based TACSO office specifically focused on identifying and including smaller local NGOs in policy and project activities. Moreover, the project funding itself also encourages networking and tutelage, since there are specific cross-border IPA programmes to encourage new axes of NGO co-operation, and other regional programmes in which NGOs from existing EU member states apply for project funding with an NGO from a candidate or potential candidates of the EU. Nonetheless, development of state and non-state capacities have been slow, and compared with its neighbours, BiH has lagged behind in the development of environmental governance. There have been modest achievements, but there is still much to do.

The next chapter will focus on a number of road-building-project EIA processes in BiH to examine the interaction between state, non-state, and international stakeholders within the legal frameworks in both entities.

3
Environmental Impact Assessment (EIA) Processes in Bosnia-Herzegovina

As mentioned in the previous chapter, the pivotal institutions responsible for environmental governance for major issues lies at the entity level, since the central state is relatively weak and there is no centralized Ministry for Environment – although the state-level Ministry for Foreign Trade and Economic Relations (MOFTER) does undertake some minimal co-ordinating role. The EIA laws in the two Bosnian entities, the Federation of BiH (FBiH) and Republika Srpska (RS), broadly adhere to the most recent version of the EU Directive (2003/35/EC), which includes the provisions of the Aarhus Convention stipulating public participation in environmental decision-making. In both entity laws, there is a requirement for interested parties to be able to access relevant information and provide feedback at all stages of the environmental decision-making process (Art. 36 of FBiH Law, Art. 35 of RS law).[1] The first step in the process is to determine whether the construction and operation of the proposed infrastructure will have a significant environmental impact. For example, there is a regulation in FBiH that lists projects with major impact for which an EIA is required, including oil refineries, motorways, chemical plants, etc.[2] For other projects, the decision to conduct an EIA will be made by the environmental ministry at the entity, cantonal or municipal level depending on the size of the potential environmental effects.[3] The stages of the process are then outlined in two phases:

> Phase I: If an EIA is needed, the project developer must submit the required information (called the Preliminary Environmental Assessment) to the responsible ministry to determine the scope of the EIA study.

Phase II: In both entities, the responsible ministry defines the scope of the EIA and instructs the developer to select a consultant to carry out the study from a list of registered local institutions. For projects deemed to have a smaller impact, cantonal (FBiH) or municipal (FBiH and RS) authorities execute the same function.

In Phase I, the developer sends the following required documentation for the Preliminary EIA to the responsible ministry: general project description; proposed measures to mitigate negative environmental impact; discussion of alternatives and the justification for the selected alternative; relevant excerpt of planning documents; data needed for the impact assessment and potential difficulties with data collection; and a non-technical summary. The responsible ministry then sends this documentation to relevant institutions and other stakeholders (including local authorities in affected communities) for comment within 30 days of receipt (Art. 58 of BiH law, Art. 59 of RS law).[4] After the 30 days have elapsed, the responsible ministry considers the comments from the relevant stakeholders to define the scope and content of the EIA for the developer (Art. 59 of FBiH law, Art. 60 of RS law).

Once the scope and content of the EIA study are defined, the developer selects one of the companies or institutions that have obtained certification for EIA to carry out the study. The developer then sends the completed draft EIA to the responsible ministry for further review and approval of the EIA study, which is Phase II of the EIA procedure. The draft EIA is sent by the ministry to relevant institutions for opinions and comments to be received within 30 days. The ministry also arranges a public meeting near the proposed site of the project, providing at least 15 days' notice of the details of the meeting (in local media and/or on the ministry website) and providing the relevant documents locally or on the ministry web pages (Art. 64–65 in RS law, Art. 62 in FBiH law). The responsible ministry must then publish minutes of the meeting promptly (within three days in FBiH, within eight days in RS), and the public can send further feedback to the ministry for a period of 30 days after the public meeting (Art. 61 in FBiH law, Art. 65 in RS law). The comments are then sent to the developer to address for the final EIA, which is resubmitted to the responsible ministry. The responsible ministry decides whether to accept or reject the final EIA study after a series of evaluation processes (Art. 64 in FBiH law, Art. 68 in RS). In addition to the requirements above, if there is a potential transboundary (either inter-entity or interstate) environmental impact from the proposed project, in line with the Espoo Convention on EIA in a

Table 3.1 Comparison of selected features of the EU Directive on EIA with laws on environmental protection in FBiH and in RS

Selected parts of EU EIA Directive	Laws in BiH
Art. 3: The EIA shall identify direct and indirect effects on humans, flora, fauna, soil, air, water, climate, material assets, and cultural heritage.	This list of requirements is included in both Laws on Environmental Protection (Art. 53 in FBiH; Art. 55 in RS)
Art. 4(2): Projects in Annex I of the Directive must have an EIA. For projects in Annex II, the authorities make a case-by-case assessment or set thresholds for an EIA.	In both entities, laws and by-laws related to environment (e.g., Law on Water Protection) include a list of projects and thresholds that are subject to EIA studies.
Art. 5(3): The developer should provide information on the details about the project, possible mitigation, environmental impact, alternatives explored, and a non-technical summary.	For the preliminary EIA (Phase I), this information should be provided to the responsible entity ministry by the developer. This information should also be included in the final EIA study (Phase II).
Art. 6: The public shall be informed by the appropriate means about the project, decision-making process, and public consultation. There should be 'early and effective' opportunities to participate in the process for deciding whether the project EIA is approved.	Interested parties within the public must be consulted during all stages of the process. During Phase I, the ministry makes documents available for comments for 30 days and organizes public hearings near the site of the proposed project. During Phase II, the ministry sends the draft EIA to relevant institutions and interested parties, so that comments can be received within 30 days. Public hearings are organized near the site of the proposed project and announced within local media and on the ministry web pages.

Note: Based on Fagan and Sircar (2010) and personal communication with a former official at FMOIT.

Transboundary Context (1991), these affected external parties should be included in the assessment process (Art. 63 in FBiH, Art. 68 in RS law).

Case studies in Bosnia-Herzegovina

The case studies used are EIA processes from road-building projects funded by multilateral international finance institutions, including the World Bank, EBRD, and EIB. The reason for choosing these larger projects is that they have higher potential for significant impact, thus a higher likelihood of conflict and necessity for co-ordination between local, national, and international stakeholders amongst state and non-state actors. A further reason for selecting these particular projects is that the investment by IFIs provides a first-hand external evaluation of the EIA processes, since they have explicit ethical commitments to the monitoring and mitigation of social and environmental impacts resulting from any of their investments contained within their own guidelines, as well as a commitment to build governance capacities. The final guiding principle for case study selection for BiH is that, due to the weakness of state authority and pivotal role of the ministries at the entity level, it is necessary to include at least one case study from each entity. In FBiH, the research examined the EIA public hearings along the Mostar Bypass, the Sarajevo Bypass, the Buna-Neum Road, and the Blagaj-Počitelj section of Corridor 5c. From RS, only the Banja Luka-Gradiška EIA is included. Although the Banja Luka-Doboj (in RS) road public consultation had been timetabled for 2010, it was not organized during the course of the research.[5] A short summary of each EIA examined is included in the following sub-sections.

Banja Luka-Gradiška Road[6]

The RS Roads Directorate published the final version of the EIA study for the building of a motorway between the Bosnian-Croatian border at Gradiška to the Banja Luka Airport (Mahovljani) in August 2004. The EIA formed part of the overall feasibility study for the road and had the following objectives: updating the EIA undertaken by the RS Urban Institute in 2000 to adhere to both EBRD guidelines and new RS legislation; identifying and addressing shortcomings in the 2000 EIA; conducting a public consultation related to the project; and drafting an Environmental Action Plan (as required by the multilateral investors).

The existing road M16 between Banja Luka and Gradiška is the busiest north-south thoroughfare in RS and forms part of the E661 on the European road network, connecting to Croatia at Gradiška (and ultimately ending in Hungary). The road south of Banja Luka goes through Jajce, Livno, and then over the Croatian border to Split. The corridor has been studied for further transport development over 40 years, starting with the preliminary study for a highway between Okučani (where Corridor 10 highway meets E661) and Banja Luka in 1973. Thus, although the road is not a part of Corridor 5c, the main internationally financed road-building project in BiH, the E661 connects to the pan-European Corridor 10 in Croatia along the main Zagreb-Belgrade motorway, and the E661 also links to the Adriatic coast. The rationale for the Banja Luka-Gradiška motorway project is to relieve congestion on the main north-south route in RS, as well as improving road transportation connections to Western Europe and to the Adriatic coast, which will lead to growth in commerce and improvement of the Bosnian economy. The highway project has been divided into three sections:

S1 – Croatian border (Gradiška) to Mahovljani (Banja Luka Airport) [29 km]
S2 – Mahovljani to Glamočani [5.8 km]
S3 – Glamočani to Banja Luka [9.5 km]

The upgrade for S3 has already been completed, and the road has been expanded into a four-lane expressway. The EBRD/EIB investment is only for S1, so the EIA and EAP documents were only produced for the 29-km stretch from Gradiška to Mahovljani. By contrast, S2 investment for the road has been provided by the RS government, and there has been no EIA, though there are plans to do so in the future.[7]

The authors of the study considered five alternative alignments for the road, which included an amended version of the 1973 preliminary design (changed due to constraints around the airport), upgrading the existing road, and a slightly modified version of the spatial plan for Gradiška. The authors of the study selected the 'West-hill' variant for two primary reasons. First, this variant minimizes the use of agricultural land and would thus have the least detrimental socio-economic impact on farming in the area. Secondly, although further mitigation measures would be needed, the 'West-hill' variant would pass mainly through uninhabited areas and would thus cause the least amount of disruption to local communities due to noise.

Sarajevo Bypass

The Sarajevo Bypass project was conceived as a means of relieving existing traffic bottlenecks in the western suburbs of Sarajevo along the M17. At present, there is a major bottleneck in this area, since there is a combination of transit traffic and local traffic, and addressing these problems would significantly reduce journey times and would improve the urban environment in the vicinity of the new road due to the reduced congestion. The completed Sarajevo Bypass would form part of the motorway along Corridor 5c (also known as E73), which goes from Budapest in

Hungary, crosses the border into Croatia (Slavonia) and through Osijek, then intersects the motorway through Corridor 10 near the town of Sredanci, before crossing the Bosnian border. As mentioned above, the Bosnian section of the road stretches between Zenica and Mostar, after passing through Doboj in RS. The road then crosses the border into the Adriatic side of Croatia, passing through Metković before ending at the Adriatic port of Ploče.

The Corridor 5c had been defined in 1981 by public agencies within the Socialist Republic of Bosnia-Herzegovina (SR BiH) and the spatial plan for the Corridor was ratified by the Republic government in 1982.[8] However, the original plans were destroyed. Initial construction of the Sarajevo Bypass had started but was halted by the war in 1992. Identification of the construction of the highway through Corridor 5c was a priority for the country's post-war poverty reduction strategy paper (PRSP) for 2004–2007 (IMF, 2004).[9] Overall, the benefits of the motorway would include: reduction of travel times, reduction of travel costs, better trade links with the Adriatic Sea and Central Europe, and employment enhancement.

The Spatial Plan from SR BiH has not been superseded by a more recent plan, so the pre-war plans remain the valid document for the design and construction of the sections of the Corridor 5c highway, including the Sarajevo Bypass. A more detailed set of parameters for the project is set out by the City of Sarajevo spatial plan (1986), and the post-war spatial plan was updated and approved by the cantonal authorities in 2000.

The proposed Sarajevo Bypass consists of the following sections:

Bypass East Branch (A1) – 4.13 km
Bypass West Branch (B1) – 3.77 km
Link road (C1) – 3.94 km
Feeder road (D) – 1.73 km

The East Branch of the Bypass (A1) branches off from the existing Zenica-Mostar road to the north and west of Sarajevo, which then connects to the Link road (C1) and the West Branch of the Bypass (B1) at the Butila exchange. The road C1 is an east-west road that connects to the existing city road network. The road B1 continues to the south-west, where it leads to a feeder road (D) that links to the existing M17 road (E73) to Mostar, then over the Croatian border to the Adriatic coast. Alternatives were considered to this configuration of building a new bypass road, connected to the City of Sarajevo by a new link road. The status quo option was not ideal, because of the traffic bottlenecks mentioned above. Widening the existing roads would be detrimental, since they pass through residential areas, and thus the change would not solve the problem of mixed transit and residential traffic, and noise and pollution would be increased. Finally, constructing the bypass without a link road would not provide an easy 'release route' to the city. The pre-war plan for the location of the interchange included a proposal for the diversion of the course of the Bosna River. The engineers of the EIA report considered a second option not requiring the diversion of the river, locating the interchange further south and avoiding sensitive vegetation, and moving further away from settlements (thus causing less noise pollution in populated areas). However, this second option would divide a floodplain and leave the area potentially susceptible to flooding. Moreover, the alternative is different from the spatial plan, and it falls outside the corridor reserved for the road in the planning documents. The designers of the report settled on a compromise solution whereby the original pre-war alignment would be retained, but instead of choosing to divert the flow of the Bosna River (with adverse hydrological effects), two bridges would be constructed.

Buna-Neum Road (Neum-Stolac section)[10]

As with the other road projects in BiH, the improvement of road transport between Buna and Neum (in Herzegovina) would also have significant ramifications for the long-term development of the area, since Neum is the only Bosnian outlet to the Adriatic Sea. However, the existing post-war boundaries are not congruent with transport and other infrastructure from the socialist period. Before the dissolution of Yugoslavia, Neum developed in the 1970s as a transit point along the Adriatic coast, between the port of Ploče (where the rail line also terminated) and the international tourist destination of Dubrovnik. However, the municipality of Neum is now essentially cut off from the rest of the country, without any modern roads connecting it with the rest of Herzegovina. The fastest route to Mostar (the main city in Herzegovina) is crossing in and out of Croatia, whilst the existing direct route is a gravel road that would not be able to handle the demands of increased tourism and economic development. Due to the current conditions and the legacies of recent history, Neum has suffered from a lack of economic development, resulting in depopulation (especially amongst younger sections of the community). The other end of this part of the road terminates in Stolac, a Herzegovinian town also affected by the legacies of the conflicts in the 1990s. The spatial planning documents from 1987 for Stolac were destroyed during the war. More recently, one of the possible alignments of Corridor 5c would have passed through Stolac, and representatives of the municipality went to the public hearings to promote the alignment through their town. However, the proposed route will pass through Čapljina instead.[11]

The road project consists of three sections:

Section 1: Old Neum-Kiševo (3.021 km)
Section 2: Kis-Broćanac (8.50 km)
Section 3: Broćanac-Drenovac (27.00 km)

At the time of the research, Section 1 of the road was already under construction, and the main phase of the project had commenced for Section 2, which means that the alignment had been finalized. For Section 3, the preliminary design phase was near completion and a variant had been chosen as the best alignment.

There will be several potential benefits to the socio-economic development of the area. As mentioned above, the existing fast route to Neum passes through Metković (in Croatia), and the completion of the new road would avoid any interstate crossings. This would fit more broadly into the long-term strategies identified by the entity government to develop the tourist sector around the coastal town of Neum and to also facilitate more rapid transport to and from Neum for the agricultural industry. When the road is completed, there will be no long waits at the Croatian border to transport goods. Moreover, since there is no existing port at Neum, there are long-term plans to build further infrastructure in Neum. This development strategy does envision the town of Neum as the main socio-economic beneficiary, but there are also benefits for other places along the route. Other villages in the Neum municipality are especially underdeveloped, but these settlements, along with villages around Stolac, would benefit from the road, since transit traffic would visit sites of cultural heritage and services along the route, hitherto bypassed by the fast route passing through Čapljina and Metković.

Mostar Bypass

The Southern Bypass road around Mostar will be a relocation of the current main road M 6.1 between Široki Brijeg and Mostar, currently aligned to pass through urban areas. However, not surprisingly, with increased traffic in and around Mostar, the combination of transit and local traffic has made it necessary to build a road that bypasses human settlements, with the exception of Rodoč, and decreases traffic congestion in the area.

The Southern Bypass has four proposed sections:

Section 1: Medjina to Miljkovica. The main design and part of the work for this section has been completed.
Section 2: Between Miljkovica and Draga Bay, which has two options – the basic and tunnel solutions.
Section 3: Starts in Draga Bay and ends in Rodoč on the southeastern slope of Hum Mountain.
Section 4: Goes through the settlement of Rodoč and connects back to the M 6.1.

For the reasons mentioned above, plans to create a Southern Bypass around Mostar were already in place 30 years ago. The previous spatial plan for BiH was adopted in 1981, and was only superseded by the spatial plan for the FBiH entity ratified at the end of 2010. The decision to create a new spatial plan at the cantonal level in the Herzegovina-Neretva Canton has been made, but the document itself has not been drafted yet. At the municipal level, the spatial plans for Mostar were only valid until 1990, and then extended to 2000. Thus, there is no current spatial plan for Mostar. With the future plans to complete the motorway along Corridor 5c, for which Mostar will be an important regional hub, it is critical to develop the road to bypass populated areas (and thus ease traffic). Although the plans had been in place, the wars in the 1990s and the lack of investment delayed the start of the project until now. Legacies of the conflict and post-conflict normalization will also have important effects on the trajectory of the Southern Bypass project. First, although there was a well-defined corridor within which the road is to be built within the expired plans, a combination of legal and illegal construction has made it necessary to alter the alignment of the proposed road in certain locations. For example, along the first section of the Southern Bypass, the alignment of the road had to change in the village of Medjina, where the local authority had issued a series of permits for businesses making it impossible to adhere to the existing plans. Second, based on maps that authors of the EIA obtained from the Mine Action Center (MAC) of BiH, there are some cleared areas but also some areas around Rodoč that are deemed 'suspicious areas'.[12]

Corridor 5c between Blagaj and Počitelj

As mentioned above, the motorway along Corridor 5c (which also includes the Sarajevo Bypass) was a part of the planning documents from the 1970s, and it is considered an important development objective in the local government's strategy for poverty reduction. The completion of the road would provide rapid road transport to the Adriatic coast to the south and to other pan-European road networks in Central and East Europe. The project for the construction of the motorway is divided into four sub-sections or LOTs. According to an issue paper drafted by CEE Bankwatch, there are unresolved issues regarding potential negative environmental impacts along LOT 3 (Sarajevo South to Mostar North) and LOT 4 (Mostar South to the Croatian border) of the highway. Part of the motorway in LOT 3 would pass through a proposed national park. In LOT 4, the highway would pass near culturally and ecologically sensitive sites around Blagaj and Počitelj. In 2007, the Commission to Preserve National Monuments in BiH[13] submitted an application to include Blagaj on the UNESCO World Heritage List, due to the natural heritage of the Buna spring (one of the largest in Europe and various architectural constructions from the Ottoman period.[14] The Commission also identified Počitelj as a candidate for inclusion as a UNESCO World Heritage site, to preserve the architectural heritage of the Old Town from the Ottoman period. This includes the Gavrankapetanović House, built during the course of the 16th and 17th centuries, and renovated in the 1970s to house an artists' colony.[15]

Both Blagaj and Počitelj are thus on the UNESCO Tentative List for protection. These areas of significant natural and cultural heritage have been identified in the EIA study for the motorway along Corridor 5c. In LOT 4, the Buna River spring and other natural features are specifically mentioned as protected areas in the study.[16] The EIA also explicitly mentions the important cultural heritage status of Blagaj and Počitelj, and proposes a series of mitigation measures to minimize the negative impact from the road.[17] These proposals include constructing the road as far away as possible from these sites; conducting tests on air, water, and soil in the vicinity of the sites; waste management; traffic management; co-operation with relevant cultural heritage institutions; and replacement of damaged land. The proposed route would be 1.7 km from the Buna River spring and thus a safe distance away with respect to environmental impact, but the visual impact affecting the environs of the spring (and the historical buildings nearby) has not been included (CEE Bankwatch, 2009a). The visual impact of a proposed bridge over the Neretva River near the Old Town of Počitelj had been raised during the EIA public consultation and initially ignored, though more recent proposals have addressed this concern (CEE Bankwatch, 2010).

State and public actors: differing trends in RS and FBiH

Since the research did not include a recent EIA public consultation from RS, it is difficult to chart the trajectory of governmental and public body capacities in RS since the changes in the EIA procedure in 2002. The EIA procedure for the section of the Banja Luka-Gradiška road was the first major project for the ministries and the entity-level roads directorate. Despite the steep learning curve, officials from the Ministry of Ecology who participated in the process were satisfied that although there was a strict legal framework with which all stakeholders gradually became acquainted, the public consultation was good and all of the steps were followed.[18] In the first phase of the procedure, ministries, governmental research institutes, and the Internet were consulted to identify the necessary stakeholders to invite for two scoping meetings in January 2004 at the RS Roads Directorate office in Banja Luka.[19] The invitees included representatives from the affected municipalities (Gradiška and Laktaši), RS ministries, university faculties, public agencies, and local NGOs.[20] The meetings were both chaired by the Director of the RS Roads Directorate.[21] Two representatives of the RS Ministry of Ecology were present at the first scoping meeting, and none attended the second meeting. To identify the problems or suggestions, the technical consultants from the French company BCEOM drafted and distributed questionnaires at the meetings, but only four were returned. The Public Consultation Disclosure Plan (PCDP) drafted at the start of the process outlined the subsequent plans for the public consultation, starting with a press conference in February 2004 to which all important RS radio, television, and newspaper outlets would be invited.

The public consultation (i.e., Phase II) would then begin with the completion of the draft EIA and would be notified accordingly in local media. For example, a notification about the public meeting for the EIA was published in *Glas Srpske* on 20 May 2004.[22] The meetings themselves were organized in the two affected municipalities by the local governments in Laktaši and Gradiška. The municipalities were responsible for arranging a meeting for local residents and participated in the process. Moreover, the co-operation between the municipal authorities and the RS Ministry of Ecology went well, and the former have become more familiar with the legal procedures associated with the EIA.[23] However, the role of both local and entity-level governmental institutions was weak overall, and more focus was given to the minimal adherence to the new legal requirements rather than understanding the impact of the road project. Some stakeholders had a suspicion that the governmental institutions (at municipal and RS levels) did not understand environmental impacts as they related to road-building.[24]

As mentioned in the previous chapter, there is little capacity for environmental governance at the municipal level in RS, and there are no specific individuals or departments at the municipal level (with the exception of Banja Luka). These capacity problems are compounded by the situation at the RS Roads Directorate. According to officials at EBRD, the co-operation between the bank and the RS Roads Directorate was particularly good during the Banja Luka-Gradiška road EIA, since there was an official at the Directorate who had specific expertise in environmental aspects of road-building.[25] Moreover, there was adequate staffing for the RS Directorate duties during the EIA process, including the hosting and chairing of the meetings.[26] Although the Environment Department at RS Roads Directorate had sufficient capacity, the incoming RS government changed the staffing at the public agency, including the removal of the aforementioned official with specific expertise on the environment and road projects. At the time of the research, there was only one official in the Environment Department of the RS Roads Directorate who was not employed at the public agency during the earlier EIA. Thus, despite the increased demands on governmental institutions during the EIA process, the capacities and experience have actually been reduced since the completion of the EIA procedure for the Banja Luka-Gradiška road.

In addition to the problems with capacity due to staffing, there are other financial challenges which open the door for conflicts of interest in the EIA process. In particular, since the Ministry is unable to afford to pay for the EIA studies, the technical consultant responsible for the study is selected by the developer through a tendering process. In other words, the developer of the project has chosen, though in a competitive process, the company that will evaluate the environmental and other impacts of the project, leading to a possible conflict of interest.[27] This potential problem with the selection of technical consultants also exists in FBiH, where the developer for the projects (the Federal Roads Directorate) has the final say in the company preparing the evaluative study. Another potential pitfall in the existing legal framework is the process of stakeholder selection. In FBiH, there is only a requirement to include 30 stakeholders, so it is up to the authors of the study as to the exact identities of the invitees to the EIA procedure. This may allow the developers to avoid organizations and agencies opposed to a project, and there is room for manipulation.²⁸ This type of strategic invitation to the consultation process is also evident in RS, where there is a suspicion by some environmental NGOs (ENGOs) that certain organizations are invited for every EIA procedure, including the public consultation of Banja Luka-Gradiška road, and the approval of EIAs in other high-profile projects, such as the thermal power plant in Stanari.²⁹

There was a similar learning process for officials in state and public bodies in FBiH. Because the legal requirements had just been instituted, officials in the FBiH Ministry for Environment and in the FBiH Roads Directorate were not used to the protocol, since public disclosure and consultation was not the norm in the former Yugoslavia. In particular, the officials at the Federal Roads Directorate had to learn that the EIA had ceased to be a process 'just about engineering'.³⁰ Moreover, the procedure was driven forward by young, qualified, and enthusiastic individuals in the Roads Directorate, and in particular, at the Ministry for Environment. For the Sarajevo Bypass project, there were six consultative meetings – three at the scoping phase and three for the draft EIA study – held in the three affected municipalities of Ilidža, Novi Grad, and Vogošća. As with the public hearings in RS, officials in the affected municipalities arranged the venue for these meetings within their locales. The EIA study for the Sarajevo Bypass was completed by Louis Berger, a French engineering firm, whereas the laws have since changed so that only companies certified by the FBiH (none of which are foreign) are able to complete EIA studies within the entity.³¹ The Roads Directorate was present at all public hearings to answer questions, and in addition to its expected role, also arranged and facilitated meetings. The reason for this is that the FBiH Ministry for Environment did not have adequate staff and office space to undertake the leading role in the EIA process, and one of the lessons that needed to be learnt from the Sarajevo Bypass consultation is that FMOIT had to become the 'competent authority'.³² However, as mentioned above, the staffing for processing environmental permits remains low, and it still takes several months to receive replies from the Ministry.³³

Nonetheless, there have been important changes at both the Roads Directorate and the Ministry. Following the Sarajevo Bypass EIA, the World Bank stipulated that the Roads Directorate establish an Environmental Department. The aforementioned official at FMOIT has now moved to the Directorate and heads the one-person Environmental Department. This official acts as a liaison between FMOIT and the Directorate, and starts the procedure for obtaining the environmental permit from FMOIT by collecting the necessary documents. This had been done by

Louis Berger and ERM, the German firm that prepared the feasibility study, for the Sarajevo Bypass project. FMOIT now has a stronger role in the public consultation process. At the public hearing for the Mostar Bypass in Rodoč on 2 September 2009, an adviser from FMOIT chaired the meeting, and an audio recording and minutes were collected by a second official from FMOIT. The team from the Ministry was joined by the Project Manager and Head of the Environment Department from the Roads Directorate. The FMOIT set the agenda for the meeting and along with the Roads Directorate officials and the author of the study from Eko Plan (Mostar) answered the citizens' questions.[34] The Chair for the meeting also oversees the technical commission that ultimately approves or rejects the EIA study.[35] The same team of officials from the Roads Directorate and FMOIT often attend public hearings for relevant road project EIAs. The same officials from the two institutions did indeed attend the two public hearings in Neum and in Stolac 12–13 January 2010 as part of the public consultation process for the Buna-Neum road EIA. The EIA had been designed by the firm Energoinvest of Sarajevo. Unlike the procedure at the start of the meeting in Rodoč for the Mostar Bypass, the Chair of the meetings in Neum and in Stolac outlined the public disclosure at the draft EIA stage of the process. The Ministry had burnt the relevant documentation onto CDs and sent three copies to each of the municipalities (i.e., Stolac and Neum), as well as sending invitations and CDs to identified stakeholders (including entity and cantonal ministries, public bodies, and NGOs). Announcements for the two public meetings were made on 21 December 2009 in the major newspaper *Večernji List* and on the FMOIT website.[36] The Chair and Head of Environmental Department at the Roads Directorate then outlined the legal framework and background information about the financing of the road project before handing over the proceedings to the technical consultant from Energoinvest.[37]

At these three public hearings, the official responsible for issuing environmental permits in the Herzegovina-Neretva Canton was present but only took the role of an interested stakeholder. At the meeting in Rodoč, the official made a comment on the provision for measures to mitigate impacts from noise, and clarification of the exclusion zone and location of the public consultation.[38] He pointed out that none of the members of the expert commission that evaluates the EIA were from either Stolac or Neum at the Neum hearing, and only introduced himself as a 'consultant' to Energoinvest at the Stolac meeting.[39] In general, the official saw his role as support to the FMOIT and provided information to FMOIT at the entity level and MOFTER at the state level for projects of mutual interest. As outlined in Chapter 2, cantons do not have the adequate capacities to manage the workload for permits and to actively participate in major EIA studies co-ordinated at the entity level. For example, the Sarajevo Canton only had involvement in the scoping phase for the Sarajevo Bypass project. FMOIT sent the scoping study to the cantonal ministry, which sent the documents to research institutes in the canton for expert opinions. Since these institutes were connected to the canton, the Sarajevo Canton Ministry did not have to pay any additional fees. These opinions from the research institutes were then sent back to FMOIT. The cantonal Planning Institute had some concerns with the impacts of the road alignment on cultural heritage and water. Additionally, one of the inspectors from Sarajevo Canton attended the meetings because of a proposal to build a concrete factory near the road, which concerned local residents. The cantonal Ministry for Environmental Planning was not invited to the second phase, did not receive any of the documents, and did not see the final permit. However, there is a high workload, so the primary concern is to process permits at the cantonal level, not participate as stakeholders in larger EIAs.[40]

In contrast to the support roles played by the Ministries for Environment in Sarajevo and Herzegovina-Neretva Cantons, the Minister for Construction and Urban Planning for Herzegovina-Neretva Canton played an active role in suggesting alternatives for the alignment of the Corridor 5c highway between Blagaj and Počitelj. Since the Minister is a trained engineer, he co-operated with the Sarajevo-based engineering firm IPSA in the development of an alternative alignment, and also took part in a roundtable discussion in Sarajevo in May 2008 during which the Minister suggested the alternative route to the Principal Designer of the Project (from Zagreb-based firm IGH).[41] Moreover, the Minister acted as a conduit between national and international NGOs that had contacted him, and local coalitions based in Blagaj and Počitelj opposed to the proposed alignment of the motorway.[42]

As mentioned in the previous chapter, municipalities in BiH do not play a substantive role in the consultation process for EIA studies, with the exception of the logistical support for the organization of the public meetings and dissemination of the relevant documents to local stakeholders. In the City of Mostar, there are two Senior Experts in Environmental Protection within the Department of Public Utilities and Environmental Protection, both of whom attended the public meeting for the Mostar Bypass in Mostar.[43] At the public meeting in Rodoč, there were three other officials who arrived late to the public hearing and identified themselves but took no further part in the meeting.[44] During the Neum public hearing, the municipal representatives underlined that they were supportive of the road project and asked for clarifications about the EIA process and funding for the road project.[45] At the Stolac public hearing, representatives of the municipality intervened by saying that they had contacted the FBiH Roads Directorate to obtain copies of the maps from the designers of the road used in the meeting (that showed the two alternative alignments), but that these documents had not been received.[46]

In all of these public meetings held in FBiH, there was little evidence of a conflict of jurisdiction between cantonal and entity-level authorities. The only exception during the hearings attended during the course of the current research occurred at the Stolac meeting. An official from the Cantonal Ministry of Agriculture, Forestry and Water Management questioned whether the EIA process had adhered to the cantonal legal framework. The Roads Directorate and FMOIT representatives of the meeting replied that the responsible stakeholder was the entity-level public agency for the Management of the Adriatic Sea and was included in the process, and that the water use and other permits would be issued after the environmental permit according to the new legal requirements.[47]

However, a more visible and ethnified conflict between and within governmental layers occurred in relation to the alignment of the Corridor 5c motorway between Blagaj and Počitelj. The original alignment in the design by the Zagreb-based IGH would pass through agricultural land, so an alternative was proposed and passed by the FBiH government that would pass through the rocky uninhabited land in the hills above Blagaj. However, this change was resisted within the entity-level government, and four ministers, all of whom are Croat, resigned in August 2009.[48] This was followed by allegations that these and other Croat politicians had purchased land along the original route and had planned to sell the land back at a high price during the expropriation process. The conflict about the alignment is not only an intra-entity dispute, but there are also disagreements between the entity and state-level officials involved in the project. A report from CEE Bankwatch outlined the arguments from the state-level Ministry for Transport and Communications (MKT). The officials from MKT did not see any need to change the alignment of the road, since they felt that the number of homes affected had been overestimated by the NGOs in Blagaj, only one large farm would be disrupted, and the visual impact in Počitelj was not severe, given the costs to build the road along the alternative route and to complete the necessary approval of new spatial plans (CEE Bankwatch, 2009b). However, the Head of the FBiH Roads Directorate suggested the alternative route above Blagaj and believed that building the bridge across the Neretva River in a different place (thus avoiding negative visual impact) would be an economically more viable situation (CEE Bankwatch, 2009b). Thus, there are fundamental differences between the state and entity-level authorities responsible for the Corridor 5c project.

In general, there are heterogeneous trends in the two entities. In RS, the only competent authority for environmental permits for major road-building projects is at the entity level. However, the staffing of the main governmental stakeholders actually decreased since the Banja Luka-Gradiška EIA, and individuals with the relevant experience left the Roads Directorate when the new government was elected. The governmental authorities in FBiH share some of the same problems, namely lack of staffing and budgetary capacities, expertise held by individuals (not institutions), and a resulting large backlog of permit applications. However, unlike the counterparts in RS, there is some evidence of a gradual change of practice in both the Roads Directorate and in the Ministry. These changes were encouraged by the international investors and technical consultants, which had a tutelage role in addition to their primary responsibilities within the EIA process. The pivotal impact of this social learning will be explored in the next section.

International actors and social learning

The EBRD has a specific mission statement to work as a multilateral lending institution (made up of approximately 60 member countries) for the purpose of transition and development, and the bank is active in many European post-communist and post-socialist states. EBRD has a particular focus on the Western Balkans, and has worked in the region since the dissolution of Yugoslavia in the 1990s.[49] The close co-ordination of the EU and EBRD has been institutionalized in the Western Balkans Investment Framework, launched in December 2009 and co-founded by EBRD, the European Commission, EIB, and Council of Europe Development Bank in order to pool resources for investment in priority areas of infrastructure development.[50] The role of EBRD and these other institutions as development-led donors means that the monies invested for various infrastructure projects do come with 'strings attached'. The requirements for the EIA process for the road projects under study were covered in the previous EBRD Environmental

Policy document of July 2003.⁵¹ The document contains specific requirements for different types of investments (based on the potential environmental impact), including the steps in the EIA process and contents of the final study. If the local legislation is less rigorous than the EBRD stipulations, then the bank's rules were to be followed above and beyond the beneficiaries laws on EIA. In the case of BiH, the newer legislation implemented during the EIA processes for the road projects in the two entities largely adhered to the European and international standards written into the EBRD document, except for the inclusion of an Environmental Action Plan (EAP), which is the same as the World Bank's Environmental Management Plan (EMP). The EAP or EMP identifies urgent environmental issues associated with the project and proposes a detailed implementation plan to address it. In the case of the investment for the motorway along Corridor 5c, the EAP is incorporated into the detailed design of the project.⁵²

However, the influence of the EBRD on the EIA is not limited to the conditionality by adhering to the 2003 Environmental Policy alone. EBRD has a physical institutional presence in BiH, with a Sarajevo office, and this made it possible for officials from EBRD based in BiH and in the head office in London to actively participate and to intervene during the EIA studies for the Sarajevo Bypass and Banja Luka-Gradiška roads. In the case of the Sarajevo Bypass, the proximity of the EBRD country mission to the FBiH Roads Directorate meant that there has been daily communication for any projects in which EBRD invests.⁵³ The bank took an active role in auditing the EIA process and examined the final document.⁵⁴ Additionally, EBRD staff based in BiH attended the scoping meetings for the Banja Luka-Gradiška road and Sarajevo Bypass, and London-based EBRD officials travelled to BiH for the draft EIA public hearings and also visited the field during the design phases of the Corridor 5c project.⁵⁵ Most importantly, EBRD attended meetings with the technical consultants, ministries, and the roads directorates at each stage of the EIA process to review the documents, check that the comments from the consultation process had been incorporated into subsequent drafts of the documentation, and advise the competent authorities on how to proceed.⁵⁶

Thus, EBRD, like the World Bank, had specific rigorous environmental stipulations for its project. By contrast, EIB has no explicit environmental policy, and thus its oversight regarding projects in BiH is different from the two other large banks. These differences are evident in the contrast between the Sarajevo Bypass and Mostar Bypass projects, the latter of which is led by the EIB. The EIB has no permanent mission presence in BiH, so representatives from the bank visit approximately every quarter, but otherwise, the EIB gives developers 'a free hand' in the implementation of projects.[57] This means that the EIB only requires projects to be compliant with the local legal framework, and it does not demand any specific stipulations across countries. Moreover, the EIB does not have a physical presence during the public consultations and no tutelage role, unlike other multilateral lenders. For example, there were no representatives from EIB at the Mostar Bypass EIA public hearing in Rodoč.[58]

In addition to the tutelage role played by the EBRD, the technical consultants also contributed to the social learning of local, competent authorities during the EIA processes. The French company BCEOM were the lead consultants for the Banja Luka-Gradiška project. Representatives from BCEOM were on hand at public meetings to answer questions, and BCEOM provided questionnaires to stakeholders in order to register feedback on the EIA study. It is difficult to obtain any further details about the role of BCEOM during the process, since the author of the study has since passed away, and the remaining engineers familiar with the project no longer work in the region.[59] For the Sarajevo Bypass, the lead consultant was Louis Berger, which worked on the overall management of the project and completed the economic/financial feasibility studies. The international team from Louis Berger consisted of five individuals: the EIA team leader, EIA expert, expert in flora and fauna, expert in noise reduction, and GIS expert.[60] Louis Berger, part of the Berger Group, has extensive experience working in the Western Balkans. In BiH, the firm has worked on other infrastructure projects, including the rehabilitation of water systems in the Una Sana Canton in the northwest of the country.[61] It also has experience in other parts of the region on transport infrastructure, including the reconstruction of the Sloboda Bridge over the Danube in Novi Sad (Serbia) completed in 2006. The company was the lead project manager for the construction of the bridge over the Sava River in Belgrade, which opened on 1 January 2012,[62] and accommodates six lanes of road traffic and two-way rail traffic.[63] Representatives from Louis Berger involved in the EIA process for the Sarajevo Bypass felt that the effective completion of the process was due to the close contact between the international consultants, FMOIT, and the Roads Directorate.[64] The tutelage role was assisted by ERM, a Frankfurt-based company that prepared the EIA study. ERM had five staff members responsible for the Sarajevo Bypass: a project manager, spatial planner, engineer, and experts on air pollution and noise.[65] ERM also has experience in managing infrastructure projects in the Western

Balkans, and worked on the Skopje Bypass road project in Macedonia. The consultants from the German company not only completed the required technical aspects of the EIA process, but also fulfilled a teaching role for the local competent authorities. EBRD expected ERM to drive the process through joint meetings with the Bank, FMOIT, and the FBiH Roads Directorate, where the international actors presented charts and diagrams of the process to ensure that local actors were learning about the EIA procedure.[66]

This strong tutorial role by consultants during the EBRD-led EIA processes can be contrasted with more recent EIB-managed road EIA processes. The law has changed in FBiH since the Sarajevo Bypass EIA, and there is now a requirement that the EIA study be prepared by a locally registered company. For the Mostar Bypass, the project is managed by the Italian firm C. Lotti and Associates, whilst the EIA study is prepared by the Mostar-based Eko Plan. For the Buna-Neum road EIA (for the Neum-Stolac section), the EIA has been prepared by the Sarajevo-based Energoinvest, whilst the project management is being handled by the UK-based Roughton International. For the public hearings examined for this research, the local EIA authors (Eko Plan and Energoinvest) presented summaries of the EIA study. It is unclear how or whether the international companies provided a tutorial role for these local companies, but they are far less visible than in the earlier processes, and there are no joint meetings between the EIB, FMOIT, Roads Directorates, and the international consultants as with the Sarajevo Bypass. Still, there is insufficient capacity amongst local companies to manage the whole process, yet there is a legal requirement for the EIA author to be a locally registered company. The result is that the feasibility studies tend to be prepared by international consultants, with a local firm responsible for the EIA study.[67] For example, in FBiH, the EIA for the Lašva-Donji Vakuf road was carried out by the local firm AGIS, though the feasibility study is the responsibility of the international firm Pöyry.[68] However, these arrangements still leave the possibility for knowledge transfer from international consultants to local firms. Thus, the early EIA processes where the investment was led by EBRD showed evidence of significant arrangements for tutelage (i.e., opportunities for EBRD and the international technical consultants to oversee and illustrate the EIA process at every stage). However, in more recent EIA processes, there is little evidence of this social learning structure, which may be the consequence of the investment being led by EIB, which is more 'hands off' about its investments as long as the project developers adhere to local legislation.

NGOs and local communities

The main transformation from the socialist-era environmental assessment process and the recently implemented EIA process is that the latter requires a more inclusive, multi-sector, deliberative approach to find solutions to environmental issues associated with infrastructure development that are least detrimental to all relevant stakeholders. In particular, the EIA process allows for hitherto excluded representatives from local and national ENGOs, as well as local citizens potentially affected by the construction, to voice their concern or support for the proposed project.

However, in practice, the role of the NGO sector seems to be reduced to that of an additional technical consultant, and there is little evidence from the EIA processes under study that environmental civil society organizations have been significant in voicing concerns of the local community, nor have they acted as conduits between the local population and governmental stakeholders.

The one exception to this has been the work of the Center for Environment (CZZS) in Banja Luka, which works on a national and regional level in networks with other large ENGOs, but has also participated in a social mobilization context during the Banja Luka-Gradiška EIA. Although CZZS is a leading, active, and well-known ENGO in RS, it was not identified in the list of NGO stakeholders by the authors of the PCDP, and thus did not automatically receive invitations and documentation for the public hearings.[69] CZZS requested documents from the Ministry and the RS Roads Directorate and attended one of the meetings during the second phase (draft EIA) in the municipality of Laktaši. The meeting was five years ago, so it was difficult to recall all of the details, but there were no other NGOs present at the meeting. The NGO became involved in the process for a variety of reasons, but particularly to support local citizens from the settlement of Jelica Brdo, who would be displaced by the construction of the road. There was also a concern that emissions from a tunnel built along the road would pose environmental health risks to the nearby community, including a kindergarten downwind from the road.[70]

Ultimately, both issues were resolved in favour of the status quo, since the organizers of the meeting posited that the public consultation covered the section of the road from Gradiška to Mahovljani (airport), whereas these problems were from another point of the route. There had not been any EIA on the other parts of the road, since they were funded by the RS government, and thus there is a potential for manipulating the input of non-state actors, first and foremost by employing 'salami tactics' to the various sub-sections of the road. In other words, the more controversial parts of the road were exempt from the scrutiny of the EIA process, which was only carried out since it was required by the EBRD.[71] The second problem, as suggested by NGOs in FBiH, is that because the stakeholders are selected by the authors of the project, it is possible to include more technical organizations instead of ones that may challenge the project and mobilize the local population. The only organization that was invited and attended the scoping meetings was Ekos, which is not a community organization, but rather a quasi-NGO with a high degree of scientific-technical capacity, since it is headed by a professor at the Faculty of Science from the University of Banja Luka. Although invited, the REC office in Banja Luka did not participate in the public hearings, since it is not within its remit to have a participatory role in domestic consultative processes because they are not an 'expert institution'.[72] Other ENGOs, such as Kanjon, also did not participate in the EIA process for the Banja Luka-Gradiška road.[73]

The lack of a community-participatory role of ENGOs was also evident during the Sarajevo Bypass EIA. Most of the invited NGOs included in the PCDP for the Sarajevo Bypass are quasi-NGOs headed by university professors that draw on the technical resources and expertise in their respective faculties.[74] As in RS, the REC Country Office in Sarajevo does not participate in project-specific public hearings, but rather works with governmental and supranational bodies during consultative policy-making processes, such as the harmonization of environmental legislation. The large Sarajevo-based NGO, Fondeko, has been involved in various international projects and other activities associated with environmental protection, but it was not involved in the public consultation for the Sarajevo Bypass EIA.[75] For the Sarajevo Bypass, CESD was asked for a *technical* opinion on a bridge that would go over a Ramsar (i.e., protected wetland) site, and it was necessary to use a cost-benefit analysis when considering all of the possible alternatives.[76] For the overall feasibility report for Corridor 5c, CESD was not an invited stakeholder but was asked for its technical opinion.[77] Another invited stakeholder, the Soil Science Institute in Sarajevo, did not have any direct role in the Sarajevo Bypass public consultation, but it did send a positive technical opinion about the overall project regarding soil use, although one of the other researchers at the Soil Science Institute did suggest an alternative methodology for land evaluation.[78] Environmental organizations in BiH that have more community or social activity do not have the technical capacity to actively engage in the consultation process. For example, Ekotim provided opinions for two sections of the Corridor 5c road. However, members of the NGO found it difficult to read the maps, although they were familiar with the area, and were only able to interpret the maps and suggest alternatives to the alignment when nearby mountains and minefields were identified.[79]

The peripheral participatory role of NGOs seems to have continued for more recent EIAs. For the Mostar Bypass EIA, only two small ENGOs, Eko Most and Eko Jasenica, were present. Eko Most did not contribute to the meeting, but said that it had sent a comment to the Ministry, though this was not indicated by FMOIT at the start of the meeting. By contrast, FMOIT stated that Eko Jasenica had sent in a written comment about the Mostar Bypass EIA public hearing. A representative from the organization reread the letter sent to FMOIT at the meeting. The main comment was on the shortcomings of the EIA in assessing the impact on flora and fauna. For the Buna-Neum road public hearings in Stolac and Neum, Novi Val and Eko Most had been invited, but representatives from neither organization attended, nor was there any indication that these NGOs submitted written comments.[80] In general, local citizens had a low opinion about the capacity and motivations of NGO representatives at the Mostar Bypass EIA, and felt that ENGOs were not engaged enough in the process.[81] Representatives from FMOIT were more critical about the role of NGOs in the process, saying that these organizations were weak, tended to not be involved in the meetings, and were a 'waste of money'.[82] However, this may be a misrepresentation of NGO motives during EIA processes in BiH. For example, Novi Val, based in Blagaj, has been invited to many EIA public meetings, but many of these are far away, so that attendance is difficult and the NGO does not have local knowledge about the ecology. Representatives from Novi Val suspect that the reason for this is that the Ministry selects stakeholders to make up numbers, in order to be legally compliant, and not necessarily to identify non-state actors who would be a good counterweight to the plans of the project developers.[83]

Members of local communities in BiH also suspect that FMOIT may be attempting to just fulfil the minimal requirements for the EIA process. Local NGOs have found it difficult to mobilize people around environmental issues, since civil society is weak and there is action only when a problem is in one's 'backyard'.[84] However, members of the local community have gradually become active in engaging with the process in various ways. Representatives from international stakeholders noted the level of public engagement during the earlier EIAs. Local citizens actively participated in the Banja Luka-Gradiška hearings, and citizens not only attended, but also asked questions at the Sarajevo Bypass meetings.[85] One of the international technical consultants noted that the participation was lively, and despite the post-war context, the turnout was high for the Sarajevo Bypass meetings, since the people were 'interested in their futures'.[86] Citizens have also become involved by using the legal opportunity mechanisms afforded by the internationally financed project. For example, since the locals feared that municipal officials were complicit in the status quo alignment of the road and were not on 'their side' regarding the Mostar Bypass EIA, one of the citizens wrote a letter of complaint to the EU Delegation in BiH.[87] However, though citizens in Rodoč were eager to attend future public hearings, there was some scepticism that their feedback would be incorporated into the final EIA, since the 'little guy' is ignored in BiH politics.[88] Sceptical local community members at the Buna-Neum hearings did not want to simply provide 'legal cloud cover' for road construction, as for hydropower plant construction projects in FBiH.[89] Despite all of the legal requirements to disseminate information about EIA processes through the FMOIT website and local newspapers, mobilization and information for EIA public meetings had been achieved through more informal channels, with the *mjesna zajednica*[90] playing a significant role. Most attendees from the local community near Rodoč found out about the meeting either through the Rodoč *mjesna zajednica* (which hosted the hearing) or by speaking with neighbours, whilst the invited stakeholders (from the canton and NGOs) received the documents automatically.[91]

In general, members of local communities were engaged, since the issues for the EIA affected them directly, though citizens often asked questions that were not directly related to environmental impact, but rather expropriation and other queries beyond the scope of the EIA meeting. There was also little evidence of multi-level co-ordination during EIA processes, with NGOs mainly providing technical support for governmental authorities, and often little co-operation between local citizens and municipal/cantonal representatives.

Mobilization around Blagaj-Počitelj: cause for optimism?

Although the case studies offered evidence of a lack of capacity and networking, and only a nascent, multi-level governance, the campaign to change the alignment of the Corridor 5c motorway between Blagaj and Počitelj shows how state, non-state, and international actors can come together around environmental issues. As mentioned above, the Minister for Construction and Urban Planning for Herzegovina-Neretva Canton played an active role in suggesting alternatives for the alignment of the Corridor 5c highway between Blagaj and Počitelj, and engaged with technical consultants, international NGOs, and local stakeholders during the dispute.

In transforming the local issue along this part of the proposed route into an international one, CZZS also played a pivotal role. The regional NGO, CEE Bankwatch, was alerted to the Blagaj/Počitelj dispute through the media.[92] CEE Bankwatch had worked with CZZS on previous occasions, and the Banja Luka-based organization identified local actors involved in the issue. Because CZZS is based in RS and it was easier to mobilize via a partner in FBiH, Ekotim became involved in the issue. As a Sarajevo-based organization, Ekotim had not received an invitation to the EIA for this part of the road (which is close to Mostar).[93] The coalition of national-regional NGOs was completed by CEKOR, a member organization of CEE Bankwatch based in Novi Sad (Serbia), and Zagreb-based Zelena akcija. This coalition of NGOs contacted various local authorities, including the cantonal Minister for Construction, who then put these international NGOs in contact with local stakeholders in Blagaj and Počitelj. To understand the issue fully, representatives of the four regional NGOs interviewed officials at the state-level Ministry for Communications and Transport, the FBiH Roads Directorate, and local activists in Blagaj and Počitelj. The information gathered in September 2009 and February 2010 was published as reports advocating a change of alignment of the road minimizing the environmental and visual impact of the highway. Moreover, an official from CZZS made a presentation at the EBRD AGM highlighting the problems along Corridor 5c, and gave a list of questions to the lead investor of the highway project.

In parallel to the activities of large national and international NGOs, at the local level, a coalition formed consisting of NGOs, private companies, and local citizens – including the ENGO Novi Val, the head of a fruit orchard, the local Association of War Invalids, a former mayor of Mostar South-east – all led by the president of an NGO for cultural heritage based in Blagaj.[94] Public meetings for Corridor 5c EIA had been held in Mostar and Čapljina, but not in Blagaj and Počitelj. Since the meetings were not held there, and people are generally only aware of events in their own local area, citizens did not know about the EIA and were absent.[95] When local citizens became aware of the social and environmental impacts of the alignment of the road, the coalition of citizens and NGOs formed to mobilize the population by spreading information about the dispute through print and electronic media, and by obtaining 5000 signatures for a petition to change the route.[96] The owner of the local Jaffa-Komerc orchards (who is part of the coalition) did attend a public meeting for the highway, but his comments were ignored and he was advised by a Ministry official to just sell his fields. Representatives from Blagaj contacted the World Bank in 2004 or 2005, but their requests were not answered.[97] The other main settlement affected by the alignment of the highway in this section of Corridor 5c is the small village of Počitelj. The local association of citizens in Počitelj became aware of the problem in April 2008, and they signed a petition and sent it to relevant institutions. Since they did so before Blagaj and have not had any direct contact with the large regional NGOs, the campaign has been done in isolation.[98]

As of mid-2010, the issue remains unresolved, and the pros and cons of the two preliminary designs are still being determined. However, one potentially positive legacy is that the multi-level coalitions that have developed around the issue may remain active to assist other local communities to mobilize around similar issues. The coalition in Blagaj has been assisting the settlement of Bijelo Polje in their campaign to seek an alternative route for the Corridor 5c highway.[99] Moreover, permanent contacts between the active citizens of Počitelj and Blagaj have grown and consolidated over the past few years.[100] Thus, there is some limited evidence of active local civil society and multi-level environmental governance gradually developing in reaction to major road-building projects.

Conclusion

The legacy of the wars in the 1990s in BiH is ethno-territorial fragmentation, leading to multiple centres of environmental governance and policy impediments. The challenge of maintaining the four levels of Bosnian institutions without the proper financial and staffing capacity is further exacerbated by the demands of transforming and implementing a robust, EU-compliant procedure for EIA studies. More than a decade after ratification of the new legislation, shortcomings with the implementation of EIA and strategic environmental assessment (SEA) processes, particularly public consultation, continues to be highlighted by the European Commission progress reports (EC, 2013a). The role of the Ministry for Environment and Roads Directorate in FBiH has shown some improvement in the capacity to lead EIA processes, whilst, on the other hand, staffing has actually been reduced in the competent authorities in RS. However, in FBiH, these gradual advances are carried by a small number of individuals, and there is no guarantee that future EIA processes for road construction will be as efficient if the 'traveling road show' of FMOIT and Roads Directorate officials leave these institutions. More worryingly, both of the early and more recent EIA processes show evidence of possible manipulation by developers and ministries in the selection of non-state actors on the basis of technical expertise, not ability to mobilize local populations, and only non-controversial sections of the project are put forward for the EIA.

At the non-state level, local NGOs remain largely absent from the process. Larger NGOs that are invited or involved tend to be quasi-NGOs based at university faculties that offer scientific-technical opinions or evaluations of the EIA study, so there is no social counterweight to the wishes of the developer and local government. Part of this is due to the manipulation mentioned above, where even active NGOs are invited to EIAs in other parts of the country, but part of the responsibility still lies with the organizations. In the case of the more recent EIA public meetings, NGOs could have acted as a conduit between government and local communities by providing technical assistance to affected communities in advance of a public hearing. On the other hand, Bosnian NGOs are more vulnerable to following priorities put forward by international donors, which focus more on creating 'developmental' or 'transactional' capacity for networking and interacting with policy makers (Fagan & Sircar, 2011). Unlike post-socialist states in CEE, there is no post-materialist class willing to financially assist these organizations through private donations or membership fees, so ENGOs find it difficult to be independent.

However, despite these difficulties, there are prospects for progress of EIA and environmental governance more generally. The legacy of the earlier EIA processes, led by EBRD, was the acclimatization of Ministry and Roads Directorate officials to a multi-level, deliberative process through social learning. The challenge for external actors is to continue these modes of tutelage, instead of past strategies of conditionality and project funding, since transformation of environmental governance has been most effective in BiH, and previously in CEE and SEE, when European NGOs and EU bodies were able to impart the 'new rules of the game' to relevant actors in post-socialist Europe. Although this has not been adopted by EIB, the notion of social learning is at the heart of the European Commission CSF programme, though its long-term results remain to be seen. Moreover, ENGOs in the Green 10, which previously had a low level of engagement in the Western Balkans due to their own priorities and the lack of credible partners in the region, have begun to implement programmes using social learning in BiH and its neighbours.

Thus, there is an arduous and expensive road ahead for the development of a truly multi-level EIA process in both entities in BiH. However, with an increase in staffing capacity, coupled with the empowerment of ENGOs and municipalities through social learning, there is a potential for BiH, despite its difficult past and challenging present, to institute new modes of governance in its EIA procedures where other post-socialist states have failed.

4
From Pariah to Partner? The Case of Serbia

Introduction[1]

The trajectory of Serbia during the first decade of the 21st century is quite remarkable: from an authoritarian state led by Slobodan Milošević blamed for much of the violence in BiH and Kosovo,[2] to a post-authoritarian state still controlled by shadowy elements of the former regime who were hiding war-crimes suspects, and finally a credible candidate for future accession into the EU. Unlike its Western Balkans neighbours, Serbia benefitted from a long legacy of state institutions and technical capacities, and largely inherited the public administration from Yugoslavia. This chapter will provide relevant background information, which will be followed by a mapping of the state and non-state capacities for multi-level environmental governance.

Despite the devolution of some authority to the republics in the 1974 constitution, significant powers remained in Serbia, and these institutions have largely remained intact since the socialist period (Clarke, 2001). On the other hand, the provinces of Vojvodina and Kosovo within Serbia were highly autonomous from the 1970s onwards, so Serbia had less formal power over its territory compared with other republics. Unlike other communist and socialist states in Europe, Yugoslavia also allowed for a degree of supervised 'self-management' in the form of municipal governance and some entrepreneurship, such that much of the responsibility for environmental management fell to local municipal institutions and enterprises, under the supervision of semi-autonomous technical institutes (which were, in turn, under the authority of relevant ministries) (Miller, 1997). Although this semi-open political system allowed for more individual contacts with the West, it did not result in the development of a national environmental movement, and there is little evidence that there was increased public participation in environmental matters, with the exception of Slovenia (Figa, 1997).

In the aftermath of the conflict in BiH and Kosovo in the 1990s, for which Serbia was widely blamed as the main perpetrator of mass military and paramilitary violence, much of the international community attention focused on addressing the immediate humanitarian crisis, including the large number of displaced persons. Less attention was given, understandably perhaps, to the longer-term, detrimental ecological effects of the war in Kosovo, Serbia, and other places in the region. To investigate the immediate environmental effects of the conflict in Kosovo, UNEP and UNHCS formed a joint Balkans Task Force (BTF) to assess how the environmental hazards in the wake of the war could potentially have negative impact on human settlements in the region. In parallel, the Regional Environment Center (REC) carried out an environmental assessment of the immediate effects of the military conflict in Yugoslavia in 1999. In addition to the ecological threat to the human population due to the actions of the Yugoslav security forces in Kosovo, including the still-present threat of human casualties from landmines, both reports indicated that there were a number of potential environmental problems due to infrastructure damaged by NATO bombing in Serbia. The BTF report identified environmental 'hot spots' in four Serbian cities that required immediate attention to ensure that there were no long-term human effects: Bor, Kragujevac, Pančevo, and Novi Sad (UNEP & UNCHS, 1999, p. 10).

The conflicts in the 1990s not only had an environmental impact but left an indelible mark on the political situation in Serbia, which will be outlined in the next section.

Politics and government of post-Milošević Serbia

The international isolation of the Milošević regime due to the involvement of FRY during the conflicts in BiH, Croatia, and Kosovo eventually emboldened a coalition of political party and civil society actors to push for change. This fusion emerged against the backdrop of NATO bombardment and NATO sanctions. Milošević had been indicted on charges of crimes against humanity by the ICTY in 1999, but by spring 2000, it seemed that Milošević and his supporters, bolstered by total control of all economic, security, and media instruments within the country, had survived opposition to the status quo (Rowland, 2000). The government sought to amend the constitution such that the existing stipulation of presidential term limits would be removed, and the office would be directly elected instead of being selected by the parliament. There are two reasons that explain these constitutional amendments: the calculation that Milošević would not be arrested and extradited to the ICTY as long as he was a sitting head of state, and the fact that with the boycott of elections in Kosovo and Montenegro compounded with a fragmented opposition an early election would consolidate his power. Despite these measures to prolong Milošević's rule, civil society movements continued to grow, including the youth movement Otpor! ('Resistance!') and Centar za slobodne izbore i demokratiju (CeSID). Significantly, these coalitions of non-state actors were coupled with the development of a united political party coalition that would contest the FRY presidential elections against Milošević (Bieber, 2003).

Two important parties within the Democratic Opposition of Serbia (DOS) that challenged the Milošević-led government were the centre-left Democratic Party (DS) and the centre-right Democratic Party of Serbia (DSS). They organized political opposition to Milošević and his control of the state. Amongst the group of Serbian intellectuals that re-founded DS, the two most prominent names are Zoran Djindjic and Vojislav Koštunica. Djindjic took over the Presidency of DS in 1994. After becoming leader, he led sustained demonstrations in Belgrade to protest the government's attempts to overturn election results (Partos, 2003). Djindjic and DS espoused a pro-Western, liberal, and reformist political agenda. Koštunica and other individuals left DS in 1992 to form DSS, a party led by Koštunica since its establishment. The high profile of DSS was not due primarily to its vote share, but rather the personal image and popularity of Koštunica. In contrast to DS, DSS is a centre-right party that is anti-communist and anti-authoritarian, and yet has electoral support among some nationalist voters.

Leading up to the Yugoslav general election in 2000, DS, DSS, and other reformist parties united to put forward Koštunica as a candidate to contest the presidential election against Milošević. After the first round, the Electoral Commission declared that no candidate had crossed the 50% threshold, triggering a run-off. Amid accusations of voting irregularities, NGOs, local citizens, and reformist-opposition political parties organized widespread demonstrations resulting in the eventual capitulation and relatively peaceful overthrow of the Milošević-led government, and a declared electoral victory for Koštunica (BBC, 2000b). Two months after the presidential election, DOS won the most seats in the Serbian parliamentary election, as a result of which Djindjic became Prime Minister (BBC, 2000a).

Although DSS and DS worked together to oust the Milošević regime as described in the previous section, the ideological differences between the two main political parties in the coalition led to a fraught relationship, and DSS pulled out of the DOS coalition in August 2001. The danger of elements of parallel security structures and organized crime elements from the Milošević era being integrated with state structures was an important destabilizing factor in the reform of Serbian politics. Djindjic was assassinated in 2003; the convicted perpetrators of the killing were linked to paramilitary units and organized crime, and probably carried out the shooting due to nationalist opposition regarding co-operation with ICTY and Djindjic's toughening stance on organized crime (BBC, 2003).

In addition to DS and DSS, the two other significant political parties in Serbia at this time were: the Socialist Party of Serbia (SPS), the political party founded by Milošević in 1990; and the Serbian Radical Party (SRS), a far-right nationalist party founded and led by Vojislav Šešelj, who is on trial at the ICTY for preparing to carry out attacks on non-Serbs in parts of Serbia, BiH, and Croatia. After Šešelj surrendered to the ICTY in 2003, Tomislav Nikolić, the Deputy Leader of SRS, became acting leader of the party (BBC, 2008b). Although his case at ICTY is still ongoing, Šešelj was released provisionally on medical grounds in November 2014, since he has been diagnosed with cancer (B92, 2014). Despite the extreme views of Šešelj's party, SRS enjoyed a significant level of support and had the most seats of any party in the National Assembly in the 2003 and 2007 elections (though it was never part of a governing coalition). After the post-Milošević era, SPS attempted to reinvent itself as a party of the Left, and has sought to become a member of Socialist International since 2008.

The subsequent presidential and parliamentary elections underlined the delicate balancing act of the main governing parties: to move forward in reform and Euro-Atlantic integration, to keep far-right nationalist parties (namely SRS) from government, and to nonetheless gather enough support amongst the substantial right-leaning Serbian electorate to remain in power (Ostojić, 2014).

The 2008 elections were particularly divisive, since they were contested in the wake of the unilateral declaration of independence of Kosovo. In particular, continued European integration would be seen by DSS and nationalist-oriented parties as abandoning the call for territorial integrity of Serbia. On the other hand, DS and other pro-European parties stressed that the only way forward in the gradual reform was through further integration (without compromising on Serbian sovereignty over

Kosovo). The DS-led outgoing government portrayed the election as a poll on Serbia's relationship with 'Europe' by agreeing to sign the SAA between Serbia and the EU during the election campaign. As expected, signing the SAA crystallized proponents' and opponents' opinions regarding further European integration. A spokesperson for DSS reacted, '[Prime Minister] Djelić's... dirty signature was on Solana's agreement for Kosovo independence... [T]he main responsibility lies at the door of [President] Boris Tadić, who gave Judas's seal to this betrayal', whilst the President of NS said that the Serbian signatories had 'disgraced the country' (Tanjug, 2008b). Despite these pronouncements, the pro-European 'For a European Serbia' coalition won more seats than any other list, and with the support of the SPS-led coalition, formed a government later in 2008.

Although he abstained from the vote on the SAA, Nikolić had broken with the Radicals and Šešelj's leadership over the relationship between the EU and Serbia. Whilst Nikolić remained pro-Russian and nationalist, he believed, unlike Šešelj, that Serbia should strengthen its ties with the EU (Gec, 2012). Nikolić ultimately formed a new political party, the Serbian Progressive Party (Srpska napredna stranka, SNS), in October 2008. Along with Nikolić, another influential politician, Aleksandar Vučić, left SRS and became the Deputy Leader of the SNS. The SNS was involved in the organization of anti-government street protests at the end of 2011 for early elections (Illmer, 2011), but ultimately, the parliamentary elections were held in May 2012. With the support of the DS-led coalition sliding (and the increased support for SNS), Boris Tadić, the incumbent President and DS leader, resigned to maximize support for his party in early presidential elections, which were held on the same day as the general election. Nikolić and SNS led the coalition 'Let's Get Serbia Moving', which obtained 24% of the popular vote and 73 seats (of which SNS won 55). In the presidential run-off, Nikolić won 49.5% of the vote with Tadic securing 47.3% of the vote (Traynor, 2012). After the presidential election, Nikolić stepped down as SNS leader, and Vučić became the acting leader of the party. SNS formed a government with the URS coalition, along with the aforementioned SPS-led coalition. The leader of SPS, Ivica Dačić, became Prime Minister, and Vučić became Deputy Prime Minister. In the following years, SNS consolidated its support domestically through anti-corruption campaigns focusing on improper privatization (De Launey, 2014), and the party pushed for early elections in 2014. The poll was held in March 2014, with the SNS and its coalition partners winning an absolute majority in the National Assembly (158 seats), with SPS and its partners winning 44 seats, and the coalition of DS only securing 19 seats (Tanjug, 2014b). Although the SNS-led coalition did not require any additional coalition partners, the post-election government again included the SPS coalition. Vučić assumed the role of Prime Minister, with Dačić as Deputy Prime Minister and Foreign Minister (Tanjug, 2014c).

The election results represented a remarkable pro-European rebranding of both Nikolić and Vučić, who had been anti-EU and strongly nationalist as leading members of the SRS, such that other leading parties had refused to form coalition governments with them after previous polls. However, both Nikolić and Vučić have distanced themselves from their Radical past, with the former making a public apology at Srebrenica for 'crimes' committed by Serbia, whilst the latter admitted that he had previously made 'political mistakes' (BBC, 2013). They were the leading figures in the Serbian government in the negotiations to normalize relations with Kosovo and are the main governing party as Serbia opens its accession negotiations with the EU. The same could be said of Dačić, whose SPS party was once controlled by Milošević. The evolution of the political fortunes serve as a microcosm of the journey taken by Serbia itself. With political pasts as intransigent outsiders and antagonists, Dačić, Nikolić, and Vučić have all distanced themselves from their nationalist roots – as Serbia itself has – in favour of a liberal, European future. The long-standing EU engagement that shepherded this evolution will be summarized in the next section.

EU engagement in Serbia

The European presence in Serbia was established soon after the post-Milošević transition in October 2000. The European Agency for Reconstruction (EAR) was established by the EU in February 2000 in the aftermath of the Kosovo conflict to provide assistance to reconstruct infrastructure in the Western Balkans. In Serbia, EAR was involved in the reconstruction of almost every sector, with the aid most often provided in the form of large contracts signed with the state (Fagan, 2010). In the first phases of the EAR mission in Serbia (and Montenegro) in 2001, the focus was more on emergency reconstruction of infrastructure necessary to sustain the population. In addition to infrastructure development, EAR was also responsible for capacity building for NGOs in Serbia through various EU assistance programmes. In 2006 and 2007, EAR administered seven different project grant schemes for NGOs located in regions within Serbia that are less developed. Implementation of the programme was managed by the Belgrade-based NGO, the European

Movement of Serbia (Fagan, 2010). The mandate for EAR in the region ended in 2008, so EAR closed all of its local operational offices at the end of the year. The primary instruments used by the EU in Serbia, summarized in the sections below, are: financial and technical assistance, and conditionality.

Financial and technical assistance

From January 2009, the EU Delegation office in Belgrade has overseen all EU programmes in Serbia. The Delegation office has administered EIDHR assistance in Serbia since 2006. The EIDHR programme had a budget of €1.2 million in 2013 from which it provided a number of project grants related to strengthening the role of civil society in the promotion of democracy, human rights, inter-group reconciliation, and political participation.[3] Since 2007, IPA replaced CARDS, ISPA, SAPARD, PHARE, and pre-accession assistance for Turkey. The allocation of IPA funding (the sum of assistance for technical assistance and for regional co-operation) totalled €178.7 million for 2013.[4] The environment and climate change sectors have become a high priority of IPA assistance. In the 2013 National Programme for Serbia, the policy area of environment, climate change, and energy comprises the largest percentage of IPA funding (21.05%) amongst the sectors targeted by the programme. As in BiH, EU strategies for the development of civil society in Serbia have moved away from building NGO project management capacity through short-term grants to an approach that also employs forms of tutelage. In particular, two strands of the CSF, TACSO and the study visits, rely on models of social learning such that representatives of NGOs receive technical assistance from local TACSO offices or insights from other representatives during study visits. The EU was also funding on-the-ground technical assistance in the form of the TACSO programme. The Serbian office was aiming to be a central location for advice and information about funding, legal stipulations, and project management for local NGOs. The country office in Serbia was structured in the same way as in other places, and there was also a Local Advisory Group working with TACSO, that includes representatives from: the Government of Serbia; USAID; the EU Delegation to Serbia; and several representatives from prominent Serbian NGOs.[5]

Conditionality

In addition to financial and technical assistance, the EU has also pursued a parallel strategy of transforming governance through extensive conditionality in the form of a series of agreements and consultative forums in which the EU promises favourable trade relations and further future integration in return for reforms. The impetus for Europeanization gathered pace after the end of the Milošević-led government. The first institutionalized step was the Consultative Task Force (CTF), established between FRY and the EU in July 2001 to examine the reforms necessary to initiate a feasibility study before negotiating the SAA.[6] After further informal discussions, it was decided to pursue more intensive formal discussions in order to position Serbia and Montenegro for the start of SAA negotiations; the resulting forum, the Enhanced Permanent Dialogue (EPD), was established in 2003. The meetings were organized along two strands. First, there were sectoral meetings to discuss certain areas that were of specific importance to the reform process leading up to SAA negotiations.[7] In the other strand of EPD meetings, relevant officials met in the plenary to discuss important general issues, such as the separation of Serbia and Montenegro, the European Partnership, the 2006 Serbian Constitution, and the annual EC Progress Reports. Since its signing in 2008, the focus of the EPD meetings has shifted to SAA implementation. For example, during the April 2010 EPD, sectoral meetings in Brussels examined implementation of the sections of the SAA dealing with transport, environment, regional development, and energy.[8] In a more recent meeting, the EPD focused on the same topics, with the discussions on environment particularly oriented towards Serbian harmonization with EU legislation in the areas of air quality, water, waste management, industrial pollution, and protection of nature, echoing the topics targeted by the IPA programme (Tanjug, 2013c). The discussions on transport were centred on the implementation of the Corridor 10 road and Rail Action Plan, the former of which is the primary case study in Serbia examined in Chapter 5.

The aforementioned European Partnership documents were not dissimilar to those that were drafted for other Western Balkans countries following the 2003 Thessaloniki European Council meeting. The first document relevant to Serbia covered Montenegro as well, and the Council considered Kosovo (under UNSCR 1244) separately. The 2004 European Partnership focused on fundamental political and economic changes, such as army, judicial, and constitutional reform, as well as measures to tackle corruption, to improve administrative capacities, to foster a market economy, to facilitate return of displaced persons, and to ensure full co-operation with the ICTY. In the environmental sector, the 2004 European Partnership mentioned the following: establishment of a Law on Environment and Environmental Protection Agency; EU-compliant EIA legislation; and creating a pollution reduction strategy

(Council of the EU, 2004b). The follow-up European Partnership document from January 2006 was updated to include the findings from the 2005 European Commission Progress Report and mainly contained the same political and economic stipulations as before, with a more explicit demand to negotiate with the interim institutions in Pristina for a settlement of the Kosovo status. In the environmental sector, the following was included in the 2006 European Partnership (in addition to the topics from the 2004 document): start construction of a hazardous waste disposal facility; start implementation of the sustainable development strategy; and prepare activities to ensure better public awareness and civil society participation in environmental matters (Council of the EU, 2006b). The European Partnership was updated once more in 2008, to reflect the situation according to the 2007 European Commission Progress Report, the separation of Serbia and Montenegro, and the adoption of IPA as the main instrument for EU assistance (superseding the CARDS programme). In the environmental sector, the document reiterated the goals from earlier documents. As with the other Western Balkan territories, regular monitoring by the EC is published in its annual progress reports. Assessments on progress are divided by the chapters of the *acquis communautaire*. For chapter 27 (Environment and Climate Change), the EU continues to highlight the slow pace of progress, citing administrative weakness, problems with implementing existing legislation, and low levels of public participation in decision-making processes (EC, 2013b).

The limits of EU conditionality in Serbia

Although there has been significant progress in Serbia, which has resulted in candidacy for accession to the EU and the opening of negotiations, the narrow focus by the EU on security issues in exchange for further integration mirrors the experience in BiH. In other words, further development of EU integration in Serbia has hinged on the fulfilment of criteria related to full co-operation with the ICTY and normalization of relations with Kosovo, rather than wholesale economic, political, and social reforms along the lines of the *acquis* (Ostojić, 2014). This is evident from the events leading to the finalization of the SAA, Serbia's candidacy for the EU, and opening accession negotiations, which are summarized below.

Despite the opposition from DSS and other parties, two key documents were signed in June 2008 (B92, Beta, & Tanjug, 2008). The first document – the SAA – reaffirmed Serbia's commitment to democracy, rule of law, legislative harmonization, a market economy, and free trade, along with 'full co-operation' with the ICTY. The other document signed with the SAA was an interim trade agreement, which was an economic incentive for undergoing the reforms contained in the SAA. It was decided that the provisions of the agreements would be 'frozen' until the September 2008 Council meeting, during which unanimous approval that Serbia was co-operating fully with the ICTY was required for the SAA and interim trade agreement to be 'unfrozen'. Contrary to some of the critiques within Serbia, there was no guarantee of either Kosovo independence or, in fact, the territorial integrity of Serbia within the documents. The wording of the SAA confirmed the status quo situation regarding the sovereignty of Kosovo *de jure*: 'This Agreement shall not apply in Kosovo which is at present under international administration pursuant to United Nations Security Council Resolution 1244 of 10 June 1999. This is without prejudice to the current status of Kosovo or the determination of its final status under that Resolution'.[9] The Serbian National Assembly ratified both documents in September 2008, with the governing coalition and LDP for, DSS and NS members against, and both SRS and Nikolić's breakaway *'Napred Srbijo!'* abstaining.

In July 2008, before the Serbian National Assembly ratified the SAA and Interim Trade Agreement, Radovan Karadžić, one of the three main war-crimes suspects whose arrests were linked to evaluations of 'full co-operation with the ICTY', was arrested in Serbia and extradited to The Hague to the ICTY to face charges related to his role during the wars in Yugoslavia in the 1990s. The Council reacted positively to the developments, and Javier Solana (Secretary General of the Council) was sure of a positive appraisal of Serbian co-operation with ICTY after speaking with its Chief Prosecutor Serge Brammertz (B92 et al., 2008). However, Serbian officials did not receive the unambiguous endorsement from the ICTY prosecutor that they had been seeking (Thorpe, 2008). Dutch Foreign Minister Maxime Verhagen vetoed unfreezing the interim trade agreement and SAA, thus vetoing the approval of the two agreements by the Council (BBC, 2008a). The blocking of the agreements by the Dutch delayed the process of EU integration in Serbia by nearly two years. After another positive report by Serge Brammertz, the European Council lifted the block on the SAA and interim trade agreement in June 2010 to start the ratification process of the SAA with Serbia (Lungescu, 2010).[10] The SAA was concluded in July 2013, and the EU completed the SAA ratification process in September 2013.

Progress of the SAA ratification process occurred in parallel to steps to declare Serbia's candidacy for the EU. The Serbian government first delivered a formal application for EU full candidate status (instead of being a potential candidate country) on 22 December 2009, even though it had not yet completed the previous step of ratification of the SAA by EU member states. The application for candidacy has the following steps: the Council of Ministers of the EU considers the application and must unanimously agree to refer the application to the European Commission; the European Commission then sends a set of questions to the applicant; and the applicant sends the replies to the questions, after which the European Commission determines whether candidate status is approved (Tanjug, 2009a). The Council of Ministers did not refer the application to the European Commission at the first opportunity, in June 2010, preferring to defer the matter until September 2010 and instead focus on the implementation of the SAA (Tanjug, 2010a). There was not unanimous support for referring the application to the European Commission, with the Netherlands, Belgium, and Germany reportedly against the initiative. In particular, the Netherlands still retained reservations about the level of Serbian co-operation with the ICTY (Večernje novosti, 2010).

With the proviso that Serbia's co-operation with the ICTY (particularly in the arrest of war-crimes suspects Ratko Mladić and Goran Hadžić) would be closely supervised, the Council for Foreign Ministers of EU member states forwarded Serbia's application for EU candidate status to the European Commission in October 2010. In November of the same year, DG Enlargement sent the questionnaire to Serbia that would determine the Commission's opinion on EU candidate status.[11] The two aforementioned war-crimes suspects were arrested and extradited in 2011. The Commission recommended that Serbia be given the status of EU Candidate Country in October 2011 (which came into force in March 2012), but made the start of EU negotiations contingent upon normalization of relations between the governments in Belgrade and Pristina.

Thus, the second key issue defined by the EU was related to the resolution of the final status of Kosovo. Belgrade had previously sought to use international legal means by bringing a case against Kosovo regarding the legality of the unilateral declaration of independence (UDI) in February 2008 by referring the case to the International Court of Justice (ICJ). In August 2008, the Foreign Minister of Serbia lodged a case to the UN to be referred to the ICJ to answer the following question: 'Is the unilateral declaration of independence by the Provisional Institutions of Self-Government of Kosovo in accordance with international law?' In July 2010, the ICJ declared that the declaration of independence was not illegal, by a vote of 10:4. Not surprisingly, the non-binding judicial opinion was met with celebrations in Pristina and with criticism in Belgrade. However, the court judgement was quite narrow in its interpretation: the actual declaration itself did not contravene any international legal standards, but the ICJ deferred on the legality of unilateral secession and international recognition. This allowed parties from both sides of the argument to adhere to their positions: the legality of the declaration will encourage more countries to recognize Kosovo; but opponents such as China and Russia can point to the ICJ silence on the legality of secession, which would allow these countries to block Kosovo's multilateral recognition, including UN membership (Ker-Lindsay, 2010).

Ultimately, despite the protestations from Belgrade, the EU integration process has determined the trajectory of Serbia's relationship with Kosovo. With Kosovo's inclusion in the process of Western Balkan EU integration as a separate entity will eventually force the issue regarding its final status. Despite some posturing and obvious domestic political costs, Serbia and its electorate have chosen a path of pragmatism in recent years, and they may need to follow a strategy, shepherded by incentives from the EU, similar to their eventual recognition of the former Yugoslav republics as independent states. After ten gruelling rounds of negotiations mediated by EU High Representative Catherine Ashton, officials from Pristina and Belgrade initialled a landmark deal in April 2013 as a significant step in normalizing relations. The two main obstacles from Belgrade were policing in the Serb-dominated northern part of Kosovo and whether Serbia could block Kosovo's entry into international organizations. However, a mutually agreed formulation on both issues was reached (Tanjug, 2013a), with the Serbian government approving the deal in a closed session three days later (Tanjug, 2013b). As soon as this agreement was reached, the last impediment for the start of negotiations had been unblocked, and the European Commission agreed to start accession negotiations with Serbia. The process of screening the chapters of the *acquis* will be completed in 2015. The initial explanatory screening for environment and climate change (chapter 27) took place in Brussels in September 2014 (Tanjug, 2014d).

Thus, the recent relationship between Serbia and the EU hinged on two issues, neither of which was directly related to the quality of the general institutional and legal reforms necessary for eventual candidacy for EU accession. The first issue was that of the apprehension and extradition of the two main war-crimes suspects, Mladić and Hadžić. Due to the scale of the violence against non-combatants and its recognition as a case of genocide, the capture of Mladić was mentioned more often. Nonetheless, Brammertz repeatedly stated that the only way for Serbia to demonstrate full co-operation with the ICTY was by arresting Mladić and Hadžić, both of whom were believed by international authorities to still be hiding in Serbia. These suspicions were eventually confirmed. Ratko Mladić had lived under an assumed name in the Banat region of northern Serbia, and was arrested during a dawn raid by Serbian Special Forces in May 2011 (BBC, 2011a). Goran Hadžić was arrested in the Fruška Gora region two months later (BBC, 2011b). Both men were subsequently transferred to ICTY in The Hague to stand trial, charged with carrying out war crimes. The second issue was the normalization of relations between Belgrade and Pristina. Once both of these outstanding issues were addressed, the next stage of EU integration – the start of accession negotiations – commenced in January 2014.

State transformation towards long-term partnership

Hence, the case of Serbia seemingly represents the successes of the long game of conditionality played by the EU. At the beginning of the process in the early 2000s, Serbia was an authoritarian country that had been widely blamed for levels of mass violence that were unparalleled in Europe after the Second World War. However, within a decade, the political discourse had largely shifted away from Serbian nationalist intransigence, towards a cautious and pragmatic European path. Most remarkably, political leaders that were ostracized from ruling coalitions because of their nationalist orientation took the reins of the Serbian government and oversaw significant steps of obtaining EU candidacy and starting accession negotiations. However, much of this progress has rested on fulfilling criteria related to transitional justice and regional stability, not wholesale changes in governance. This relatively rapid progress (compared with BiH) is further buoyed by existing technical and administrative structures around Belgrade. Much of the public administration from the post-1945 Yugoslav federation (and earlier incarnations of Yugoslavia) was centralized in Belgrade, so the challenge is far less daunting than that in neighbouring BiH, where the EU and other international agencies have been involved in building the state from scratch. The state structure in Serbia is far simpler than the complex configuration of the post-1995 institutions in BiH, since Serbia is a centralized state with some limited competencies for municipalities and the autonomous province of Vojvodina. It is instructive to now turn to the institutional configuration of Serbia to map the environmental capacities of governmental and non-governmental actors, before focusing on multi-level Serbian EIA processes in the next chapter.

State competencies for environmental governance

The Serbian European Integration Office (SEIO) was established by the Serbian government in March 2004 to co-ordinate European integration in a number of ways, and its primary role is to prepare 'for adoption the acts of the Government designed for supervising, directing and harmonizing the activities of ministries and special organizations in relation to the European Union association and accession and formulates public information on the process of European Union accession and association'.[12] The SEIO is ultimately under the supervision of the Prime Minister and Deputy Prime Minister of Serbia, although it is not a department within the Office of the PM, as it is in BiH.

There are eight sections within SEIO:[13]

- Department for Co-ordination of the Accession Process and Monitoring of the Stabilisation and Association Agreement
- Department for Planning, Programming, Monitoring and Reporting on EU Funds and Development Aid
- Department for Cross-Border and Transnational Co-operation Programmes
- Department for Translation Co-ordination
- Department for Communication and Training
- Department for Legal and Financial Affairs
- Group supporting the Head of the Operating Structure and the national body responsible for IPA II
- Group for Implementing and Monitoring Implementation of EU-Funded Projects

The SEIO provides oversight in the areas of legislative harmonization, which can be illustrated by examining its role in environmental protection. This policy sector remains one of the most challenging areas of Europeanization, not only because it will be financially costly, but also due to the fact that the consequences of the environmental *acquis* traverse sectors, and would require a high level of inter-sectoral and state/non-state partnerships. In June 2014, a State Secretary from the Ministry for Agriculture and Environmental Protection estimated that it would cost €10.5 billion to achieve full harmonization in the area of environment (chapter 27 of the *acquis*) (Delegation of the EU to the

Republic of Serbia, 2014). The inclusion of the SEIO is integral to the drafting of new legislation. Before a draft law can be considered as a government proposal and sent to the parliament for discussion, it must obtain approval from three sources: from the Secretary for Legal Issues that the draft law is in line with the domestic legal framework; from the Ministry for Finance regarding the financial aspects; and from the SEIO to ensure that the draft law adheres to EU best practice.[14] Although it would have been ideal to have more time for oversight by the SEIO on the compliance of draft legislation, there were often tight deadlines (often ten days or less) in practice, so there had not been enough time to make an article-by-article analysis matching the Serbian legislation with the relevant EU directives.[15] It is only now, when Serbia's application for EU candidate status has been accepted, that there will be a thorough screening of domestic legislation, and it is only then that there will be an accurate picture of the level of harmonization.[16]

One of the most crucial tasks carried out by the SEIO is the drafting of the National Programme for Integration (NPI) of Serbia into the EU. The first NPI was drafted in October 2008 and then updated in 2009, with input from the European Commission (EC, 2008b). The Serbian NPI document had been created in order to take an inventory of all of the government activities associated with European integration so that the activities associated with eventual accession could be effectively co-ordinated. The Serbian NPI fed into the National Programme for the Adoption of the Acquis (NPAA) once the Republic of Serbia was deemed a candidate country for accession by the EU. The Serbian NPAA was published in 2013, and laid out the strategy for the harmonization of the legal framework in the country for 2013–2016 using the Copenhagen criteria: political criteria; economic criteria; and the ability to assume the obligations of EU membership (SEIO, 2013).

At the heart of environmental governance in Serbia is the state-level Ministry for Environment and Spatial Planning. The competence for environmental protection had been the responsibility of a Directorate within the Ministry for Science and Technology until May 2007, when the Ministry for Environment was established. Not surprisingly, the capacities of the newly formed ministry were quite low, and there was little evidence of inter-Ministry co-ordination with the Ministry of Agriculture, Forestry, and Water on matters of common concern (EC, 2007b, p. 34). However, within the first year of its establishment, there was evidence of significant improvement in its capacities, especially in the number of staff members in departments related to horizontal legislation and European integration in the newly formed Ministry (EC,

2007b, p. 42). Moreover, there has been some marked progress in the relationship between the Ministry and the civil society sector. The Ministry for Environment after the 2008 general election was seemingly more open to co-operation with NGOs. The Ministry applied for membership to the IUCN. As a member, the Ministry would be an 'equal partner' along with other state and non-state actors that are also members of IUCN.[17] The Minister for Environment attended the launch of a nationwide bike ride to address issues related to waste management, and has commenced discussions to sign a memorandum of understanding with NGOs regarding participation in environmental decision-making.[18] In general, leading environmental NGOs found the Minister at the time, Oliver Dulić, to be more proactive and open to dialogue than his predecessors.[19] However, there was also concern that there is no institutionalized contact between NGOs and the Ministry, and that the more 'open' relationship was down to an individual and was not systemic.[20] In other words, there was no guarantee that the positive dynamic would continue when the Minister would leave office.[21]

Despite some possible concerns with the long-term continuity of the proactive policies towards the environment, there was rapid improvement in the harmonization of environmental regulations. In May 2009, the Serbian government drafted and passed a suite of laws in the area of environmental protection designed to adhere closely to the related EU directives, whilst other legislation with the set of laws ratified international conventions (Ministry of Environment of Serbia, 2009).[22] There have also been developments in the institutionalization of European standards in environmental protection. In 1985, Serbia ratified both the Vienna Convention for the Protection of the Ozone Layer (1985) and the Montreal Protocol on Substances that Deplete the Ozone Layer (last amended in 1999) and established an Ozone Office within the Ministry to ensure compliance with EU and international standards (EC, 2009b, p. 43). In December 2011, Serbia ratified its Strategy for the Implementation of the Aarhus Convention. In the area of climate change, the Republic of Serbia ratified the Kyoto Protocol in 2008, and it has since established a Designated National Authority (within the Climate Change Unit at the Ministry) to approve projects under the Clean Development Mechanism of the Protocol, by which developed countries can implement one of these approved projects to help with sustainable development in Serbia in exchange for carbon credits.[23] However, the Ministry recognized that legislative harmonization is occurring without the necessary administrative capacities in place, and that budgetary restrictions in the near future will exacerbate the situation.²⁴ There was significant pressure from the EU to rapidly harmonize legislation such that there was not adequate time to complete an economic impact assessment to plan the steps necessary to implement the new legislation and how EU requirements will be co-ordinated across sectors.²⁵

The task of implementing environmental policies is concentrated in the Serbian Environmental Protection Agency (SEPA). SEPA was established in 2004 within the Ministry for Environment and Spatial Planning, and it had 30 members of staff (including the Director) as of 2010 (Krunic-Lazic, 2010). The Agency has the following functions: the development and maintenance of the national information system of environmental protection (including the register of polluters); collection of environmental data in order to publish progress reports on the state of the environment; and co-operation with the European Environmental Agency (EEA) and the European environment information and observation network (EIONET).²⁶ Serbia has recently established a national node on EIONET, which includes 23 domestic agencies that manage environmental monitoring data (EC, 2013b, p. 54). SEPA has been gradually gaining capacities and has compiled and maintained a register of polluters in Serbia (EC, 2009b, p. 43). SEPA is responsible for the country's European Pollutant Release and Transfer Register (E-PRTR) database, which is a Europe-wide register of environmental data from industrial facilities. In 2012, 197 of the 270 operators with an obligation to do so are included in the database, with the gaps mainly from the agricultural and mining sectors (EC, 2013b, p. 54). For these reasons, the NPAA includes the objective of strengthening 'institutional capacities through recruitment and training of human resources in jobs that involve further development, improvement and maintenance of the National Registry of pollution sources' (SEIO, 2013, p. 544).

Another environmental policy instrument established in 2004 was the Environmental Protection Fund, which was to be used for funding or co-funding initiatives in areas related to environmental protection. The revenue for the Fund was from the annual budget, monies from privatization, grants, and other sources.²⁷ The Fund had been active in the co-financing of international projects, especially in the area of monitoring air quality, sanitation, and waste management (EC, 2008b, p. 43). However, the level of funding was low, with only 0.4% GDP earmarked for the Fund (EC, 2008b, p. 43). Although there had been plans to expand the staffing of the Fund, it ceased in October 2012 and its function was not reassigned within the Serbian administrative set-up (EC, 2013b).

The final significant environmental policy institution at the state level is the Institute for Nature Protection for the Republic of Serbia, which was first established in 1948 as a research institute for the protection of natural resources. The Institute's jurisdiction is nominally the whole of the country, including the autonomous provinces of Kosovo and Vojvodina, with the suspension of Belgrade authority in the former since 1999. In March 2010, the Government of Serbia restructured the Institute for Nature Protection to create a separate Institute for Vojvodina, thus reducing the jurisdiction of the existing institution to Serbia. In addition to the head office in Belgrade, there is a working unit in the southern town of Niš. The Institute is funded directly from the national budget and is thus independent of the Ministry, but it works closely with the Ministry and with SEPA.[28] The primary function of the Institute is to provide technical expertise; to achieve this, there are natural scientists, biological scientists, and engineers employed at the Institute across a number of relevant disciplines.[29] In other words, the Institute for Nature Protection provides a 'scientific service' for the formulation, implementation, and monitoring of environmental policy in Serbia.[30]

Due to the weak administrative capacities at SEPA and other governmental institutions, and the significant task of harmonization, the 2013 EC Progress Report acknowledged that 'significant additional efforts' are needed to harmonize environmental legislation, and alongside low levels of institutional capacity, 'inconsistencies and incompleteness in legislation that prevent effective enforcement' at the state level (EC, 2013b, p. 56).

The autonomous province of Vojvodina

According to the 2006 Serbian Constitution, in addition to the state and municipal levels, there is an intermediate level of limited autonomy in the provinces of Kosovo and Vojvodina. The complexities of Kosovo disputed sovereignty have been briefly outlined above. The status of Vojvodina is less disputed, but there is still uneasiness amongst some Serbian policy makers about the robustness of the link between Belgrade and Novi Sad. Vojvodina joined the Kingdom of Serbia in 1918 at the close of the First World War as the Austro-Hungarian Empire collapsed. Before that, some or all of the territory in present-day Vojvodina had been part of the Habsburg Empire since the 18th century. Vojvodina has been an autonomous province within Serbia since the end of the Second World War. The autonomy of Vojvodina was first reduced by Milošević in 1989–1990 before it was increased after 2000, but to a lower level than during the 1970s and 1980s. However, there is no significant constitutional or extra-legal secessionist movement in the province, as there was in Kosovo, or indeed, in Montenegro. The geographic and historical particularities of Vojvodina have led to a level of ethno-linguistic heterogeneity not evident in other parts of Serbia. There are six official languages in use by official organs of the province: Serbian (Cyrillic script), Hungarian, Slovak, Croatian, Romanian, and Ruthenian.[31]

As an autonomous province, there are governing powers transferred from Belgrade to Vojvodina. In other words, there is no original power in the provincial structures, so the Republic of Serbia can unilaterally derogate from the devolution arrangements. Vojvodina has an Executive Council, which is comprised of the heads of a number of Secretariats (that are similar to ministries), including a Secretariat for Environmental Protection. A broader set of competencies of Vojvodina were clarified in the 'Omnibus Law' of December 2001, including the transferred power of environmental protection.[32] A new statute for Vojvodina was ratified by the Assembly of AP Vojvodina on 14 October 2008. However, opposition parties in the provincial Assembly objected to the provisions of the statute, claiming that the powers suggested within the document would be unconstitutional and would promote separatism, since the proposed statute would promote the Executive Council into a government, and there would be the possibility of sending regional representatives to Brussels. The opposition proposed 155 amendments to alter the statute, yet all but one was included in the approved draft (B92, 2008). The draft statute was then referred to the state-level National Assembly for approval, where it met similar opposition. The opposition was concerned that too much authority would be ceded to the province, which would be the competent authority in 20 sectors, including environmental protection, and approximately 150 new powers overall (Tanjug, 2010b). Some members of the opposing benches evoked past agreements that led to eventual secession, such as the 1974 Yugoslav Constitution, the Kumanovo Agreement (ceasing the conflict between NATO and FRY), and the Ahtisaari Plan (to settle the status of Kosovo) (B92, Beta, & Tanjug, 2009). Nonetheless, the National Assembly, with support from government parties and the opposition LDP, approved the statute, though the DSS-NS opposition coalition filed a motion with the Constitutional Court in December 2009 challenging the constitutionality of the statute (Tanjug, 2009b). Nonetheless, the then President Tadić brought the statute into force by decree in January 2010.

In the area of environmental protection, the division of labour between state and province is relatively clear in the constitutional arrangements, though there is some confusion about the competence of Republic and Provincial inspectors for projects on the territory of Vojvodina. A clearer set of regulations had been under discussion.[33] If the new statute for Vojvodina had been fully implemented, there would be four new areas of competence: environmental inspections for installations on the territory of Vojvodina; establishment of a Provincial Institute for Nature Protection; the return of authority of Fruška Gora; and establishment of a fund and separate budget.[34] The Executive Council would be called the Government, which would mean that Secretariats would be renamed as Ministries, but this would be cosmetic (i.e. ,the name change would not affect the substantive competences of provincial institutions for environmental governance).[35] However, in July 2012, the Constitutional Court deemed a number of the provisions in the statute as unconstitutional, and ruled that autonomy in many sectors, including environmental, agricultural, and rural development policies, should be curtailed (Lynch, 2012). Furthermore, the Court ruled in December 2013 that two-thirds of the statute were in contravention of the Serbian Constitution. A six-month deadline was set for the adoption of an amended statute, which was adopted by the Serbian National Assembly and Assembly of Vojvodina in May 2014. The new statute explicitly deemed Vojvodina to be an 'inseparable' part of Serbia (Tanjug, 2014a), and the level of autonomy is greatly reduced compared with the previous statute (Beta & BIRN, 2014).

Thus, in light of the secession of the other republics and the gradual international recognition of Kosovo, Vojvodina's position within Serbia remains a divisive issue.

Municipalities

The most local levels of self-government in the Republic of Serbia are the municipalities and cities. There are 167 city or municipal units in Serbia, excluding the City of Pristina and 28 municipalities in Kosovo.[36] The larger, local self-government units are designated as cities, of which there are 23 within Serbia. Moreover, some of the cities are sub-divided further into 'urban municipalities'. The cities with 'urban municipalities' are: Belgrade, Niš, Kragujevac, Požarevac, and Novi Sad. For example, Belgrade has 17 such urban municipalities, which have limited powers of self-government compared to municipalities outside Belgrade. The competent authority for policy processes related to environmental governance lie at the city level, and smaller sub-units within Belgrade do not carry out procedures such as EIA.[37] The centralization of

From Pariah to Partner? The Case of Serbia 111

environmental governance is similar in Niš, the main city in Southern Serbia. The urban municipality of Crveni Krst in Niš has no authority over environmental matters, and only contributes occasionally for specific problems, such as problems with ambrosia (i.e., ragweed).[38]

The lack of administrative capacities for governance at the local level in all policy sectors has been indicated in the EC Progress Reports as an ongoing problem, with a gap in staffing and technical expertise at the municipal level, significant disparities in capacities between municipal governments, and the urgent need for further training and professional development (EC, 2013b, p. 8). Much of the problem has to do with the lack of qualified individuals to undertake tasks related to environmental governance in some of the smaller towns.[39] The municipalities that are 'industry-intensive' and are building infrastructure that may affect the environment are of particular interest, and the government has started to provide some training for these municipalities.[40] However, there is a significant gulf to bridge between localities with high levels of human and financial resources and areas in which the situation is more difficult, particularly areas in Sandžak and South-east Serbia. Both state and non-state actors have had successful co-ordination with municipalities, both in urban and non-urban settings. The municipality of Prijepolje is located in Sandžak. Although it is in one of the less wealthy parts of Serbia, the municipality of Prijepolje is quite proactive in proposing initiatives related to environmental protection and has opened a local 'Green Office' to collect citizens' ideas about conservation.[41] Within Belgrade, the municipality of Vračar has had an Adviser for Environment since the early 2000s, and focuses on NGO co-operation, improving green spaces, and education.[42] The municipal authorities have seemingly been more transparent with NGO funding and less prone to clientelism compared to other municipalities.[43] Another outlying location with a positive stance towards ecology is Vranje, a municipality in Southern Serbia close to the border with Kosovo, which banned the use of plastic carrier bags[44] with local government approval in 2010. Although the authorities in the town of Dimitrovgrad co-ordinated effectively with the developer during the EIA for the Corridor 10 highway, other smaller authorities in Southern Serbia co-operated well, including Vranje and Leskovac, and even the small municipality of Bela Palanka.[45]

However, other interactions between municipalities and environmental governance actors have been less positive. For example, NGOs had approached the council for environmental protection at the municipality of Stari Grad in Belgrade, but the ruling political party refused, perhaps due to the concern that NGOs would interfere.[46] The Green

Network of Vojvodina, a well-established NGO based in Novi Sad working on sustainable development, has extensive contacts with the relevant secretariats at the provincial level, but found it difficult to meet with officials at the city level, despite several attempts to contact officials working on the topic.[47] Another prominent NGO, Young Researchers of Serbia, had been involved in a project funded by EAR in 2006, in which the public utility company and local government of the New Belgrade municipality were partners. The NGO found it hard to interact with both municipal partners, since the operating procedures had not changed from the previous era, and there was a general unwillingness to work with civil society organizations.[48]

The Standing Committee for Cities and Municipalities (SKGO), a non-state body, is the primary organization that handles the co-ordination of activities amongst towns and municipalities. The SKGO is a member organization in which the members are the 167 towns and municipalities in Serbia.[49] The origin of the inter-municipal body stretches back to the 1950s, but the current form of the SKGO has only been operational since 2000.[50] The role of the SKGO is to represent the interests of local self-government to state-level authorities, to provide training to municipalities about the existing legal framework, and to find areas of common interest amongst municipalities.[51] The SKGO also organizes several committees for policy areas for which inter-municipal co-ordination and communication are crucial. For each committee, the Chairperson is elected in the General Assembly by other municipalities. The remaining members of the committees are selected by SKGO officials from the relevant departments. For the SKGO Committee for Environmental Protection, officials at the Department for Communal Issues, Urban Planning and Environment make the selection. The choices are made such that there are representatives from urban and rural locations, and there is regional diversity. The representative is either a mayor, vice-mayor, representative from the environmental department (if it exists), or municipal expert for environmental protection.[52] There are approximately 30 members of the Committee for Environmental Protection.[53] The municipality of Batočina chaired the Committee for Environmental Protection during the research period in 2010, as it had done in the previous session.[54] The first organizational meeting to agree on the committee agenda and constitution for the session had been arranged for Vrnjačka Banja, so that municipal officials would be able to share information about their new waste management system; but ultimately, the meeting was moved to Veliko Gradište, a municipality in Eastern Serbia along the Danube and on the Romanian border.

This first sessional meeting for the Committee for Environmental Protection, observed during the research, opened with a presentation by the Mayor of Veliko Gradište about the Danube as the 'main street of Europe', and how the development of environmental protection would promote local tourism, using the example of Srebrno jezero (Silver Lake) in Veliko Gradište.[55] After introductory remarks by an official from the SKGO Department for Communal Issues, Urban Planning and Environment and the newly elected President of the SKGO, as well as brief introductions by the municipal representatives, much of the meeting consisted of rapid confirmation of procedures and documents for the Committee. There was little debate, and confirmation was done through a show of hands. Although most of the representatives were not active during the meeting, there were a few delegates who did participate. Upon nomination for Deputy President of the Committee, a representative from Niš underlined that it would be important for the Committee representatives to think about how to apply for EU IPA funding and further co-operation with local NGOs.[56] The representative also invited the other members of the Committee to Niš for a fair on sustainable development and environment, and a donor conference, to demonstrate an example of good practice by municipal governments in the area of environmental protection.[57] Niš is the largest city in Southern Serbia, and a university town, so it is unsurprising that the representative was proactive. However, individuals from non-urban municipalities also actively participated in the committee meeting. Koceljeva is a small non-urban municipality in North-west Serbia, yet the representative at the committee meeting suggested that the municipalities specifically organize roundtables events for NGOs, which would also improve links between local self-government and NGOs.[58] During the meeting, the committee also appointed representatives for a working group that would examine the legal framework and propose a way to implement the laws regarding the use of natural resources that would not be detrimental to municipalities. The first volunteer for this working group was from Bajina Bašta, a small rural municipality in Western Serbia that is by the Drina River on the Bosnian border near Tara National Park. The delegate from Bajina Bašta came forward, since the municipality, though quite small, had governance experience with the use of local water and forest resources.[59]

The representative from Bajina Bašta also indicated that there was no reliable audit of the capacities of committee members, so it would be useful to compile a database of human and technical capacities for environmental governance.[60] Although there was some interest at the meeting, the initiative was ultimately not pursued by SKGO. However, the SKGO assisted us in administering a questionnaire of existing capacities and experience with international donors, by sending the survey via e-mail to all municipalities, and then following up the query thrice. There were ultimately 45 responses, with variation in size, region, and type of local self-government (i.e., city, municipality or urban municipality).[61] These data were complemented with face-to-face interviews with officials with competence in environmental governance in Belgrade and Novi Sad.

Unsurprisingly, urban settlements seemed to have more institutionalized arrangements for local environmental governance. The best example of this is Novi Sad, where the diversity of the population means that all official documents issued from the city must be available in multiple languages: Serbian, Hungarian, Slovak, Romanian, and Ruthenian. Thus, any public announcement must be published in daily newspapers printed in those languages.[62] At the time of the research, the City Administration consisted of 16 permanent employees and four additional members of staff on a one-year contract as part of a graduate scheme.[63] All of these employees were involved in the process of issuing environmental permits, and six of these members of staff (i.e., one co-ordinator and five other individuals) worked specifically on EIA processes.[64] Moreover, Novi Sad is a member of the SKGO Committee for Environmental Protection, and the City Administration for Environmental Protection had a budget of 88 million RSD.[65] The Head of the City Administration for Environmental Protection would have preferred more specialization amongst staff members to deal with specific sectors (e.g., chemical industry), but the existing structures were capable of managing the workload.[66] The City of Novi Sad mainly dealt with environmental permit applications related to petrol stations and mobile telephone masts. The latter type of installation has become more frequent since 2008, when the Austrian company VIP has been allowed to operate in Serbia.[67] Novi Sad, the second-largest city in Serbia, is in a stronger position than smaller towns in Serbia, since it is a university town and can thus draw on a wider pool of potential candidates with experience for processes such as EIA.[68] The administration in Novi Sad may have a higher level of capacity in Belgrade, the largest city in Serbia. At the time of the research, there were ten members of staff working on EIA processes in the City of Belgrade, though there was quite a heavy workload such that it was necessary to take work home to complete all of the applications. Thus, it would have been ideal to have more members of staff to handle the number of applications for environmental permits.[69] Nonetheless, Belgrade does have a separate section within the administration dealing with environmental issues. Additionally, all of the cities that responded to the survey have a separate department for environmental protection.[70] The cities are located in different parts of Serbia, including Vojvodina (Novi Sad and Pančevo), Northern Serbia (Smederevo), Western Serbia (Šabac), and Central Serbia (Kraljevo, Kruševac, and Kragujevac). However, smaller municipalities and even urban municipalities with few areas of competence have separate departments that address environmental issues. For example, the urban municipality of Surčin, west of the centre of Belgrade, has little responsibility over environmental matters, since almost all processes are centralized at the city level. Still, the municipality has a separate Department for Economy, Agriculture and Environmental protection with four employees and a budget of 4.5 million RSD in 2009.[71] Mali Idjoš is a Hungarian-majority municipality in Vojvodina with 13,494 people (according to the 2002 census), and there were no secondary-school students or graduates within the municipality in 2006–2007 (Statistical Office of the Republic of Serbia, 2009). Mali Idjoš nonetheless had a small Department for Commerce, Agriculture, Communal Management and Environmental Protection with two officials focusing on environmental protection and a modest budget of 1.9 million RSD in 2009. Although it is difficult to establish whether the survey has a representative sample of local self-government, it found that, of the respondents: the urban municipalities of Obrenovac and Crveni Krst did not have any officials addressing the environment; all of the cities did have such a department, as did six other municipalities, and the urban municipalities of Zemun and Surčin; and the remainder of the municipalities had one or more officials dealing with environmental protection. However, in many of these cases, one or more of these designated officials worked simultaneously on other sectors in the municipal administration, such as agriculture (Bačka Topola), development (Bajina Bašta), spatial planning (Majdanpek), and forestry (Rača). There were also several examples of municipalities in which these officials were also working as inspectors, such as Bela Palanka, Bojnik, Mionica, and Topola. This presents a potential conflict of interests between inspection and implementation of environmental governance, but in many of these municipalities, there are simply not enough qualified personnel or budget for adequate staffing in line with standards of best practice.

The questionnaire data also revealed the extent to which respondent municipalities have been involved in the building of multi-level environmental governance. With the exception of Novi Sad (which has other strategy documents), all of the cities surveyed had drafted and ratified a LEAP. Whilst cities such as Smederevo and Kragujevac relied on domestic or local government funding, other cities such as Pančevo and Kraljevo funded the LEAP through international donors, the EAR/ Danish government, and SDC, respectively. Smaller towns and communities also had or were drafting LEAP documents, many of which were funded by international donors such as SIDA (Trstenik and Bajina Bašta), REC (Vrnjačka Banja), EAR (Majdanpek), and UNEP (Bor). The urban municipalities of Barajevo, Medijana, and Obrenovac had also ratified or were in the process of creating a LEAP. The other 22 urban municipalities and municipalities did not have a LEAP.[72]

An important finding from the questionnaires was that approximately half of the local governments surveyed had received some form of international assistance since 2004. In rare cases, this help has been the form of technical assistance, such as the participation by Smederevo, REC guidance for the drafting of the Vrnjačka Banja LEAP, and UN Habitat training for Kraljevo in 2008. However, for the most part, international assistance was given in the form of large, infrastructure projects. Several of the respondents mentioned waste management project money from the Gesellschaft für Technische Zusammenarbeit (GTZ), a publicly owned German enterprise specializing in sustainable development. For example, Beočin received funding for a landfill from GTZ in 2009. The Western Serbian municipality of Kosjerić received over 1.5 million RSD for containers to dispose waste, whilst the Vojvodina town of Opovo secured GTZ funding in 2005 and 2009 for a dump truck and development of a local landfill, respectively. Other donors have also funded similar initiatives. In 2007, UNDP invested approximately €50,000 in Bojnik for the purchase of containers and lorries for rubbish transport. USAID provided approximately six million RSD for waste collection and management in 2006. Other infrastructure funding originates from governments in EU member states. For example, the municipal government of Ravenna in Italy provided funding for the establishment and maintenance of air quality monitors, and the Italian Ministry for Health and Ecology funded a project addressing air quality, water quality, and waste management in Pančevo.[73] The Rača municipality received around 35,000 RSD from the South Moravian regional government in the Czech Republic for the construction of a PET recycling plant. The same region has also funded the botanical garden and repair of the wastewater treatment plant in Kragujevac. The responses suggest that, until now, there has been little technical assistance and social learning from international donors. The emphasis on projects instead of other forms of assistance is not dissimilar to the situation in BiH. The problems with such an approach are not lost on affected local actors, who realize that there are many projects funded by international donors, yet there is a lack of a co-ordinated, long-term strategy for systemic improvements.[74]

Despite the 'projectization' of development by donors, there are at least two examples of programmes that focus more on technical assistance. The first was the MEGA programme formulated and implemented by USAID, which ran between 2005 and 2010. The municipalities were selected on criteria along five axes related to local economic development (LED): general characteristics (population around 70,000 persons, importance as a regional centre, proximity to important natural feature or transport); level of co-operation (responsiveness to application procedures, interest in LED practice, co-operation with agencies involved in LED); commitment to LED (resources and staff earmarked for municipal LED, and initiation of LED processes); brownfield and greenfield investment opportunities; and stability of the ruling local coalition.[75] The criteria were designed to not result in selection bias regarding region, nor were urban municipalities overrepresented. Willingness to co-operate was an important factor in the selection process.[76] There were 32 municipalities that participated in the MEGA programme. As with the EU-funded MISP Programme (which started in 2003), local administrations were generally willing to be helped, since they want access to European and international grants and loans.[77] The assistance for MEGA included several technical assistance workshops in areas such as EIA, feasibility studies, and preparation of infrastructure projects.[78] Although there are some municipalities that still have low levels of capacity, there have been improvements in some MEGA municipalities: there is good inter-municipal co-operation amongst MEGA municipalities; there are officers responsible for the environment in all MEGA municipalities; and municipal staff has been educated in environmental matters.[79] When the new suite of environmental laws was under consideration, MEGA and the Chamber of Commerce organized public meetings, and MEGA and the EU helped with the drafting of the Law on Construction.[80] Thus, there have been significant improvements using a bottom-up approach of training and capacity building around environmental governance for MEGA municipalities.[81]

In addition to the MEGA programme, the series of study visits organized by the Belgrade-based NGO European Movement in Serbia (EPUS) also addressed issues around environmental protection. EPUS is an active NGO with a high level of financial capacity that has worked to provide training, education, and networking to governance actors in the area of European integration, and the NGO was the implementing partner for a number of EAR projects. The final activity of the EPUS programme was to organize two stages of study visits for mayors and city managers from 12 Serbian municipalities. The first group went to Brussels for a seven-day visit, whilst the second group went to Newcastle, Cardiff, and Belfast to see how these cities managed EU funds and focused on environmental protection. The result of the visits was that the municipal representatives returned home and were motivated to adapt models of best practice from cities in EU member states.[82] Since the completion of the programme, some participating municipalities have been successful in securing external funding for projects in the environmental sector: Kruševac received a grant of €3 million for water management, and Čačak received €2 million of investment for a landfill, and Paraćin also received monies for waste management.[83]

Thus, there is a gradual development of municipal environmental governance capacities, despite continued concern by the EU reflected in the annual EC Progress Reports. The questionnaires suggested the gradual development of institutionalization of environmental governance. The respondent cities all had separate departments, as well as a number of municipalities, and even a few urban municipalities. Of the respondents to the survey, the only local administrations that did not have any individual or department responsible for environmental protection were two of the urban municipalities. Most of the international donor assistance afforded to the municipalities was in the form of large project grants for environmental infrastructure (e.g., wastewater treatment plants) from EAR, UNDP, GTZ, and others. Few programmes used social learning or tutelage as the focus of the assistance, but when these methods were used, in the form of study visits or training sessions, they were effective in the development of environmental governance capacity.

ENGOs in Serbia

Before analysing the current status of ENGOs in Serbia, it is instructive to flag up two points: citizen distrust towards all NGOs, and the weakness of ENGOs compared to NGOs in sectors that developed at the time of the post-Yugoslav transition in the 1990s.

In most post-authoritarian states, the legacy of the old regime is a deep level of distrust amongst citizens towards the NGO sector, a pattern that is even more acute in post-Milošević Serbia. Due to the surge of foreign donor investment in the NGO sector starting in 1990s and continuing through the start of the 2000s, most local citizens feel that NGOs are wealthy and simply 'money laundering machines' (OSCE, 2010, p. 40), or agents of unwanted Western influence,[84] though most international donors have started to focus their funding in other parts of the world. Even though NGOs tend to be less well-resourced compared to earlier times, the suspicion towards these civil society actors continues for two reasons: lack of response to salient issues and lack of media attention (OSCE, 2010, p. 41). Most NGOs have also found it difficult to address salient issues as they come to the fore, since most organizations follow the priorities set by donors in order to receive funding. In other words, priorities are set from the outside by donors, not from within the society by local communities. Moreover, many NGOs continue to lack support and co-operative relationships with media, since the latter were seen as being agents of the previous regime and not providing an independent voice. Although this negative image still exists amongst many communities, especially away from Serbian cities, there are some encouraging indications. In particular, if there are any questions in areas such as displaced persons or human rights, a representative from the government interviewed during the research admitted that they would approach NGOs for an opinion and that media outlets do the same.[85] Moreover, the area of environmental protection in particular is not a divisive issue amongst local communities, unlike other issue areas on which NGOs focus, such as transitional justice.[86] Since the environment is not contentious and scores highly on surveys of citizens' priorities for their own community, it is an obvious site for grass-roots mobilization.[87] Because of the public interest in environmental protection, it has been easier for media to name and shame polluters and for this to be an effective deterrent.[88]

Despite some of the benefits for NGOs working in the environmental protection, the work by ENGOs has been made more difficult by the focus of the international donor community on other branches within the civil society sector, namely organizations working on human rights, democratization, and minority rights. The reason for this is that international donors funded democratization and human rights NGOs as agents of change in the 1990s to affect a transformation of the regime, and this was followed more recently by international funding targeted at post-conflict peace-building.[89] As a result, environmental organizations are less developed and less politically powerful compared to their counterparts in human rights.[90] Although many of the NGOs (as in environmental protection) are in Novi Sad and Belgrade, there are also organizations with significant capacities in Western and Southern

Serbia. For example, the Centre for Human Rights located in Niš has been able to provide protection for victims of torture and other forms for human rights violations. The organization Urban In is based in the Sandžak region that is home to a significant Bosniak/Muslim population. Urban In has been successful in implementing projects on issues related to tolerance and multiculturalism. These and other human rights NGOs continue to grow, but they are growing in new directions, such as children's, women's and disabled rights.[91] Environmental organizations are not as strong, though they are getting stronger gradually. One impediment is that with the head start in administrative and financial capacities, human rights and democratization NGOs have been able to slightly alter their objectives to fit most calls for grant proposals. The result has been that even when the Government of Serbia declared a specific focus on 'the Year of the Environment', organizations that have been building up their capacities in other topics have learned how to write project proposals, and they have secured the funding for environmental projects. In other words, if an NGO can write proposals, it can obtain project grant funding on any topic.[92] Another problem with this type of donor funding is that it encourages organizations to chase grants, irrespective of the topic, and there is little opportunity to build up a specialization.[93] However, some of the larger NGOs are able to resist this pursuit of donor funding and focus instead on specific areas, such as the Young Researchers of Serbia specialization in biodiversity. However, at the time of the research, with the exception of the six to seven large ENGOs, no environmental civil society organization in Serbia has the luxury to pursue such a strategy.

In fact, the distribution of ENGOs in Serbia is not dissimilar to the situation in BiH. There are only a handful of large ENGOs that have the financial and project management capacities to not only survive but also to interact with governmental institutions, and to network with other NGOs locally, nationally, and transnationally. Although there are methodological difficulties to measure the overall capacities of Serbian ENGOs using the REC 2006 questionnaire, the REC publication nonetheless offers insights into the state of play amongst Serbian environmental civil society organizations. In the REC survey of Serbian NGOs, there were 114 respondents. The budgetary data indicated a large gap between organizations with financial capacities and those without significant financial resources. Nearly one-third of the REC respondents (32.8%) had an annual budget of €1000 or no budget at all, whilst only four organizations had budgets of €100,000 or higher. However, even this distribution may not be accurate. Because of the lack of funding mentioned above due to donor flight, the number of ENGOs with the highest levels of funding has decreased, and the figures themselves may be misleading. The responses to the survey are self-assessed, so some ENGOs only receive a certain amount of money through grants for one to two years, and this is not representative of the actual level of financial capacity. In fact, many of the NGOs from the 2006 directory are no longer active.[94]

A related difficulty in gauging the number of active and effective ENGOs is the fact that not all of the major NGOs are 'real'. In other words, these organizations are more akin to consultancies rather than social movement organizations, and the orientations of these GONGOs sit, sometimes uncomfortably, between governmental and non-governmental sectors. Two of the most visible ENGOs in Serbia fall into this category. One such environmental organization is Ambasadori odrzivog razvoja i životne sredine (Environmental Ambassadors for Sustainable Development), which was established in 2004 in Belgrade. It has been very successful in obtaining grants from international donors and implementing projects with partners within Serbia (e.g., BCIF) and throughout the region. Environmental Ambassadors conducts workshops for local government and is well connected to governmental structures more generally. One reason for this is that the President of the organization, Dr Andjelka Mihajlov, was a member of the Cabinet as Minister for Environmental Protection and Natural Resources in the DOS coalition government in the first elections after the ousting of Milošević. Mihajlov would potentially be able to return to government if the opportunity arose again.[95] However, international donors and NGOs identified Environmental Ambassadors as one of the more effective environmental organizations in Serbia.[96] The other visible GONGO in Serbia is Ecotopia, which has only been a registered organization since 2009, but it has become one of the most recognizable environmental organizations in the country. Despite its recent establishment, Ecotopia has been able to complete several high-profile campaigns with the involvement of celebrities and government officials, which has been noted by international NGOs such as the Institute for Sustainable Communities (ISC).[97] Ecotopia started with a coming together of important political and economic figures in Serbia who sought to adapt models of best practice from Europe related to public awareness raising, networking, and partnerships amongst private firms, NGOs, and government.[98]

Serbian ENGOs, like their counterparts in BiH, also have few formalized links with pan-European, environmental non-state actors in the 'Green 10'. Moreover, WWF activities within South-east Europe were co-ordinated by a representative based in Belgrade.[99] Compared with the situation in BiH, some of the more high-profile ENGOs had their roots in anti-establishment, pro-democratic activism. The Green Network of Vojvodina developed out of the anti-Milošević Pančevo Peace Movement in the 1990s. After the end of the Milošević-led regime, the activists in Vojvodina became more focused on environmental issues, particularly local sustainable development.[100] The Young Researchers of Serbia started as activists during the Milošević era. The Ecological Movement of Novi Sad also had its roots in a social movement critical of the Milošević regime.[101] Thus, there are substantive links to social activism during the socialist period.

Moreover, some of the more active ENGOs have developed links and roles within a transnational ENGO community. This is not to say that there are no examples of significant participatory campaigns by active ENGOs in Serbia. Serbian Green Youth is affiliated with other organizations as a full member of the Federation of Young European Greens (FYEG) in Brussels. However, it has worked with local residents in mobilizing support for the preservation of 'Peti Park' when developers had planned to build on the park site, and Serbian Green Youth was the only on-the-ground NGO concerned with this issue.[102] The Ecological Movement of Novi Sad is involved in local direct action. For example, one of the volunteers for the organization became active in the Ecological Movement after the NGO intervened when local residents protested the building of a mobile phone mast near their homes. The Ecological Movement advised the local residents on channels through which they could complain and how to organize their campaign.

CEKOR, based in Subotica (near the Hungarian border) carries out project work for one of the Green 10 NGOs, CEE Bankwatch. It not only works on projects within Serbia, but also co-ordinates co-operation amongst non-state actors in activities across the region. For example, CEKOR worked with Zelena akcija (Green Action, Zagreb), Ekotim (Sarajevo), and CZZS (Banja Luka) to liaise with citizens and political officials in Blagaj and Mostar who were against the alignment of the Corridor 5c highway (see Chapter 3), to draft a report on possible alternative alignments of the road, and to present their findings to the EBRD.[103] Although Serbian ENGOs tend to not be involved in networks connected to the Green 10, they are active in other regional networks. CEEweb is a regional network of NGOs in CEE and SEE that focuses on biodiversity and the preservation of nature. Of the ENGOs interviewed for the research, MIS, Eko Centar, and Ecolibri Bionet are all CEEweb members. Another regional network for sustainable development, the

Northern Alliance for Sustainability (ANPED), brings together NGOs to focus on sustainable development. MIS is a member of ANPED. Perhaps the best example of transnational activity is also carried out by MIS activities on a European level. MIS received its first European funding in 2001, co-funded by DG Environment and ReREP, to develop a regional environmental NGO network, along with Zelena akcija, CZZS, and the Ecologists' Movement of Macedonia.[104] Since then, MIS has been an active member of the Environmental Forum (formerly the New NGO Forum), a consultative body composed of ENGOs from the Western Balkans and Turkey, which is funded by CARDS, that meets in Brussels to discuss priorities for donor funding and environmental policy.[105] Thus, active Serbian ENGOs have tended to be more involved with media and networking, which are more transactional, instead of more directly participatory ones. Moreover, these activities, as in BiH, tend to be fostered by international donor funding.

However, some donor activities have sought to develop more transactional activities amongst smaller community organizations that have been more focused on participatory actions. USAID funded a small project grant programme, implemented by the Balkans Community Initiatives Fund (BCIF) office in Belgrade. The call for grant proposals resulted in the funding of 25 local initiatives led by small community organizations or ENGOs. Of these 25, BCIF has continued to work with a core of twelve organizations to form a network of local environmental organizations, the Green List of Serbia, across the country.[106] Of the USAID-funded organizations that are part of the Green List of Serbia, many cite direct action as the most important or one of the more important activities, including Zeleni putokazi, Zeleni ključ, GM Optimist, Moravski orašak, Šumatovac, Green Bike, Josef Pančić, Zelena zona, Lisac, and Bela breza.[107] Green List members had various levels of financial capacity, with moderate annual budgets of €10,000–50,000, including Green Bike, Zeleni putokazi, and Zeleni ključ, whilst other organizations had more modest resources, such as Moravski orašak, Zelena zona, and Šumatovac (€1000–5000). All of the Green List NGOs that participated in the survey reported that there were numerous training sessions, especially connected to the two Green List campaigns in 2009 and 2010. In 2009, the NGO network, co-ordinated by BCIF, cleaned illegal landfills in 11 Serbian towns and participated in a 'green' charity cycle ride across the country to educate local populations about ecological protection. In 2010, Green List ran a campaign on recycling. The Green List ENGOs also reported that they have had contact with other organizations in the network through the training and project activities, and that these links have, in some cases, led to co-operation on local campaigns. However, the long-term effects of the USAID funding are difficult to gauge. As with larger ENGOs, only one of the Green List organizations have been involved in the public consultation processes for EIA studies; Moravski orašak indicated that it had been involved to provide an opinion. However, a few of the ENGOs, such as Lisac, were involved in other local, multi-level policy processes, namely the drafting and implementation of LEAP. The evidence is mixed regarding the development of transnational or transregional networks of local ENGOs. Organizations that had existing transnational links, such as Josef Pančić and Šumatovac within the scouting movement, have been able to maintain links outside Serbia, though these links are not specifically environmental in focus. However, GM Optimist applied as a partner with the Slovene NGO Zvik for IPA funding, and Green Bike has co-operated with an organization from Srebrenica (in BiH) for an IPA application to develop a transborder cycle path. On the other hand, Zeleni putokazi, Zelena zona, and Bela breza do not have any contacts with organizations outside Serbia. Thus, the USAID-funded project to create the Green List of Serbia, through networking and training and other transactional activities, has had some positive outcomes, but it is difficult to ascertain whether it has led to a long-term increase in the capacities of these ENGOs.

Conclusion

The situation in post-socialist Serbia differs from the context in BiH. Most importantly, although there were (and are) lasting effects in the country from the wars in the early 1990s and the conflict in Kosovo at the end of the decade, the main task for the international community has not been peace-building or peace consolidation, but rather the rehabilitation of authoritarian state structures and parallel extra-legal security structures into more liberal democratic configurations. Thus, the crucial difference is that BiH is undergoing a process of state creation (i.e., the building of new state structures where there were no such competencies or human capacities before the war), whilst Serbia is experiencing a state transformation. In other words, much of the infrastructure and expertise is in place, but the task of European and other international actors is to steer the state in a more 'European' direction.

In this way, the case of Serbia is not dissimilar to its CEE counterparts. All of the difficulties with the creation of an independent environmental civil society in post-socialist countries that joined the EU in 2004 and 2007 are also relevant in Serbia. However, unlike these post-socialist states that are now within the EU, the process of Euro-Atlantic integration includes a set of transitional justice criteria, such as the apprehension of war-crimes suspects, and the disputed sovereignty of Kosovo (on which there is not unanimous support within the EU). These national issues are often conflated with the whole process of state transformation on the path to Europeanization, such that attention is often diverted from the financially arduous path of institutional harmonization.

That being said, because of the centralization of structures and expertise in Serbia during the socialist period, Belgrade has a head start on the process compared to BiH. In fact, Serbia has travelled far enough along the path of integration to open negotiations for EU accession, despite the risks that the aforementioned issues related to the legacies of the conflicts would derail the process. Serbia has several ENGOs that developed as part of the democratic anti-Milošević movement in the 1990s, and some even before that, so that there is a strong foundation within the non-state sector to act as a counterweight to the government. Moreover, there is a significantly higher level of administrative and financial capacities with the Serbian state than in BiH, and governing authority is not ethnified and is far more centralized in Serbia. However, as in the CEE countries, the disadvantage of having this level of centralization is that there is little scope for new modes of governance at the local level, and municipalities in Serbia still have low levels of capacity, despite the attention of USAID and the SKGO. Although there will be challenges in the years ahead to achieve further Europeanization, the existing capacities in Serbia in all sectors suggest that the path will be less difficult than in BiH.

With these details about the state of multi-level environmental governance in mind, it is instructive to examine EIA processes in Serbia to see these processes in action, which will be the focus of the next chapter.

5
Environmental Impact Assessment (EIA) Processes in Serbia

Unlike the case of Bosnia-Herzegovina (BiH), there is a long and well-established institutional legacy of environmental governance, and governance more generally, in Serbia. As outlined in the previous chapter, the governmental structures in Serbia are highly centralized at the state level, with minimal responsibilities at the municipal level and no intermediate level of governance. The exception to this is the autonomous province of Vojvodina with limited autonomy in selected areas, though these competencies were curtailed in the amended statute in May 2014. Because of existing governance capacities, the objective of Europeanization in Serbia was not one of state consolidation and creation, but rather to transform or rehabilitate authoritarian modes of government into practices that adhere to European democratic standards. In the period after the end of the Milošević-led regime, a number of reforms were passed, including new environmental legislation.

The 2004 Law on Environmental Impact (135/04) is broadly congruent with the EU directives on environmental impact assessment (EIA), including the provisions of the Aarhus Convention.[1] The Serbian EIA process consists of three phases to determine: whether an EIA is necessary (Phase I); the scope and content of the EIA (Phase II); and the EIA study (Phase III). The legislation places a strong commitment by the developer to complete an EIA study if prescribed by the competent authorities (e.g., the state-level Ministry of Environment and Spatial Planning), without which implementation of the project cannot commence (Article 5). The developer sends an initial document with project details to the responsible authority to determine whether an EIA is required (Article 8), with a decision by the responsible ministry within 15 days (Article 10).[2] Once it is determined that an EIA is required, the developer submits a scoping study, which includes: information about the developer, project location, project description, main alternatives, summary of environmental impacts, likely significant negative impacts, proposed mitigation measures, non-technical summary of data, and potential difficulties in collecting data relevant to the study (Article 12).[3] The responsible authority then has ten days to inform interested parties – including affected communities, organizations, and governmental bodies – about the scoping study in order to collect comments; these interested parties have a further 15 days to send feedback to the responsible authority. The responsible authority issues a decision on the scope and content of the study ten days after the feedback period is over (and notifies the developer within three days of making the decision), taking under consideration the details of the project and the comments received from interested parties (Article 14). The developer has one year to send a draft EIA after receiving the final decision on the scope and content of the study (Article 16).[4]

Once the developer submits the draft EIA, the responsible authority has seven days to inform the project developer and interested parties about the time and venue for a public hearing to introduce and discuss the EIA, which must take place at least 20 days after the public is informed about the meeting (Article 20). The EIA study is evaluated by a Technical Commission, formed within ten days of receipt of the relevant documents, which is comprised of experts who do not have a conflict of interest and are not already involved in the evaluation of the project (Article 22).[5] The Technical Commission has 30 days to provide a recommendation of approval on the EIA study, which does not include periods where the Technical Commission or responsible authority requests the developer to make amendments before further consideration (Article 23). The competent authority has ten days to adopt the decision on the EIA study after receiving the final report of the Technical Commission (Article 24), and ten days from adopting the decision to inform interested parties about the decision (Article 25). If the EIA study is approved, the developer has two years within which the project implementation must be started (Article 28). In addition to details about the steps of the EIA procedure, the legislation specifies the media in which interested parties are informed: publication in at least one local newspaper for each local language spoken in the area in which the project is proposed; direct written notification by the competent authority for relevant organizations and agencies; and electronic media (Article 29). There is also a stipulation in the Serbian legislation, in line with the Espoo Convention on EIA in a Transboundary Context, that if the project has potential impacts across international borders, the foreign interested parties would also be notified about the EIA procedure (Article 32). Thus, the Serbian law on EIA appears to adhere to the relevant EU directives, which is summarized in Table 5.1.

Case studies

As with the case studies in BiH, the research in Serbia focused on major road-building projects with funding from international financial institutions (IFIs), including the World Bank, EIB, and EBRD. The reason for examining these cases is that the size of the project and international involvement provide the best conditions for testing the robustness of the policy implementation of consultative, multi-level environmental governance in Serbia. Secondly, the in-house requirements of the IFIs ensure that there is detailed documentation of the EIA processes over the sections of the road-building projects for which the IFI is the lead

Table 5.1 Comparison of selected features of the EU Directive on EIA with laws on environmental protection in Serbia

Selected parts of EU EIA Directive	Law in Serbia
Art. 3: The EIA shall identify direct and indirect effects on humans, flora, fauna, soil, air, water, climate, material assets, and cultural heritage.	The definition of EIA as it is used in the context of the legislation contains the list of required impacts to consider (Article 2).
Art. 4(2): Projects in Annex I of the Directive must have an EIA. For projects in Annex II, the authorities make a case-by-case assessment or set thresholds for an EIA.	Listed in 'Regulation on laying down of the list of projects for which the impact assessments is obligatory (List 1) and list of projects for which the impact assessment can be required (List II)' (Official Gazette of the Republic of Serbia, 114/08)
Art. 5(3): The developer should provide information on the details about the project, possible mitigation, environmental impact, alternatives explored, and a non-technical summary.	This information is included by the developer to determine the scope and content of the EIA (Article 12) and for the EIA study itself (Article 17).
Art. 6: The public shall be informed by the appropriate means about the project, decision-making process, and public consultation. There should be 'early and effective' opportunities to participate in the process for deciding whether the project EIA is approved.	The public remains informed about the EIA procedure at every stage (Articles 10, 14, 20, and 25). The means by which information is provided to the public and other interested parties is outlined in Article 29.

investor. The second reason is that these projects offer the best opportunity to find the points of resonance and of tension amongst local, national, and international actors relevant to environmental governance. The Serbian case studies focused on various road-building projects along different parts of the pan-European Corridor 10: the Belgrade Bypass; Corridor 10b (between the Hungarian border and Novi Sad); Corridor 10c or road E-80 (between Niš and the Bulgarian border); and Corridor 10d or road E-75 (between Grabovnica and the Macedonian border). Various environmental assessment plans have been implemented along the route for these sections of the corridor, but none of the public hearings occurred during the research period. Thus, all of the analysis of the public consultations in this chapter will be done using the EIA studies and interviews.

Corridor 10b

The length of the section of the corridor is 118 km and would be a vital transport link between the two largest cities in Serbia, Belgrade and Novi Sad, and Central and Eastern Europe. The northern section of Corridor 10 ends in Budapest, where it links to Corridor 4 (ending in Germany) and Corridor 5 (extending between the Adriatic coast and Ukraine). Unlike the other road-building projects considered in this chapter, the funding for Corridor 10b was provided from the Serbian budget, unlike the other roads funded through IFI loans (Government of the Republic of Serbia, 2010). As a result, the EIA process for this section of the motorway adheres to the newer post-2004 legislation yet was not bound by the stipulations of IFI investment like the Belgrade Bypass, road E-75, and road E-80.

Corridor 10c[6]

The branch of Corridor 10 that connects Niš, the largest city in Southern Serbia and third-most populous city in the country, with Bulgaria is an important route. This part of the corridor links with other pan-European corridors in Sofia to Istanbul, Thessaloniki, and the Black Sea. The completed road will hopefully contribute to the development of Niš as a regional transport and economic hub, as well as providing long-term economic development for other towns along E-80, including Pirot, Bela Palanka, and Dimitrovgrad (p. 38–39). Along with other road construction in the country and in the region, there will be better connections between South-eastern Serbia and other parts of Serbia (via Niš), including the hitherto isolated Danube-Timok region in Eastern Serbia. The new motorway will also allow transit traffic to bypass settlements along the existing road, thus easing congestion in these places and improving the quality of life for citizens. The proposed alignment of the road also has positive impacts for local attractions, allowing easier access to spa towns in the region (e.g., Niška Banja), and natural/recreational areas (e.g., Stara planina and Jelašnička gorge) (p. 16).

The road-building project for this part of the pan-European road E-80 has been divided into five sub-sections:[7]

Prosek – Crvena Reka (22.5 km)
Crvena Reka – (10.3 km)
Čiflik – Stanicenje – Pirot (29 km)
Pirot – beginning of the bypass at Dimitrovgrad (12.8 km)
The northern bypass of Dimitrovgrad (8.7 km)

Corridor 10d[8]

The road links the southern regional centre of Niš with Macedonia, eventually ending on the coastline of North-west Greece. The construction of the road is an important part of a stated national strategy to improve the road network, which had suffered from years of neglect due to economic and political conditions in the country. Although there are certain environmental and spatial planning negative impacts in constructing the road, the route would provide a vital lifeline for the depressed communities in the southern part of the country bordering Macedonia. In particular, communication and commerce would improve for the less economically developed urban areas around the road, namely Vladičin Han, Vranje, Vranjska Banja, and Bujanovac (PEIA E-75, p. 22). Moreover, the construction of the road will provide an efficient transport link to the main urban centres in the region, thus allowing for the development of commuter settlements and a service economy along the road, including petrol stations and shops, which would require local employees (PEIA E-75, Ibid.). Moreover, the accessibility of previously stagnating and marginalized communities in the region would slow migration to the urban centre of Niš, thus easing crowding in cities as well.

The preliminary design for the road is divided into five sections:[9]

Grabovnica – Grdelica (8 km)
Grdelica – Caričina Dolina (12 km)
Caričina Dolina – Vladičin Han (14.3 km)
Vladičin Han – Donji Neradovac (12.8 km)
Donji Neradovac – Levosoje (8.7 km)

Belgrade Bypass[10]

The severe traffic congestion in Belgrade and the level of pollutants in the city centre may have long-term health implications. The construction of a subway system (which is planning to open its first stations in 2018) (BIRN, 2013) and completion of the Belgrade Bypass road are amongst the most important proposals for improving the current situation suggested by city officials (Andric, 2010).

The bypass road has three sections:

Section A (9.7 km) – Batajnica-Dobanovci
Section B (37.3 km) – Dobanovci-Bubanj Potok
Section C (22 km) – Bubanj Potok-Starčevo

Of these parts of the road, Section B crosses the Sava River and Section C crosses the Danube River. Due to the aforementioned long-standing traffic congestion in Belgrade, the construction of a bypass road in the west of the city linking the E-75 and E-80 roads had been targeted in the 1980s and 1990s, with Section B funded by IFIs. However, during the conflicts in former Yugoslavia in the 1990s, any further implementation was stopped, only to be reconsidered at the start of the 21st century (CEKOR, 2007). The EIA for the Belgrade Bypass draws on previous studies from the 1990s and in 2000, as well as the City of Belgrade Master Plan (valid until 2021) that was adopted by the City of Belgrade in 2003.[11] The bypass road is being constructed through agricultural lands west of the city, with tunnels to avoid severing existing communities. Environmental assessments were carried out for sections A and B of the road, but since there were two alternative alignments for Section C, there was only a preliminary assessment of the effects of the final section of the road. The two variants of Section C are between Vinča and Bubanj Potok, west of the Danube River. One alignment (Masterplan + BPT) would require a tunnel through Bubanj Potok hill, whilst the second variant of the road (Masterplan – BPT) would not require construction of a tunnel. The detailed design of the alignment along Section C is further complicated by property expropriation and the identification of eight sites of cultural/archaeological importance identified in the EIA study in 2005.

At the time of the EIA in 2005, a two-lane road existed along part of Section B, and other structures for future tunnels and bridges had already been constructed. Thus, the immediate implementation of the project would focus on the phased upgrade of the road in Section B and construction of the road in Section A, with plans for Section C to be implemented in the future.

International financial institutions: funding and knowledge transfer

As in BiH, the IFIs involved in providing loans for the Corridor 10 project in Serbia fulfilled a dual role, both as investor and in a potential tutelage role for technical consultants, governmental representatives, and developers not familiar with international standards for environmental assessment. According to the developer of the highway project, the three IFIs – EBRD, EIB, and the World Bank – met in Luxembourg to discuss the way to divide up the investment and the assessment procedures to use for the sub-sections of the road. As decided in the Project Appraisal document, the three banks agreed to use the World Bank procedure.[12]

For this reason, the World Bank has a lead role in certain aspects of the design of the project. The organization responsible for the design of the road was determined through a tendering process by the developers. The World Bank was required to approve the terms of reference by ensuring that they adhered to the World Bank's requirements and then reported back to the developer. If there had been a problem, the World Bank would have notified the developer directly.[13] Since the World Bank rules were used for the Corridor 10 project, its rules for procedural precedence were applied during the EIA. In other words, international rules took precedence over the Serbian legislation. If the World Bank had been 'ruthless', it would only apply the international standards, but this was not the case.[14] However, as outlined below, this meant that the developers and technical consultants had to produce additional documentation in order to fulfil the IFI requirements. When the Serbian developer complained about the additional cost, the IFIs felt that it was 'not their problem', since it was necessary to follow the rules for environmental assessment to the letter in order to secure the loans.[15] The World Bank provided substantial advice and helped the Roads of Serbia public company complete this additional step of creating an EMP, but nonetheless, officials at Roads of Serbia conceded that completing these documents in addition to the EIA study was 'difficult'. Much of the interaction was between the World Bank and the Serbian technical consultants who wrote the EIA/EMP studies.[16] Although the second member of the environment department did meet the officials from EBRD and the World Bank during the EIA procedure, most of the meetings were of a higher level with the head of the environmental department (who later moved to K10) and the designers from the Highway Institute.[17] The World Bank implicitly builds in the tutelage role into its projects in Serbia, and the region in general. When it reviews project documents from developers, it provides feedback on ways to improve the procedures. In this way, the World Bank seeks to encourage long-term capacity building through its projects.[18]

The EBRD deferred to the World Bank procedure for the environmental assessment of the project, since the standards would adhere to the more stringent policy from 2008, which includes stipulations for assessing social impacts, compared to the 2003 document that focuses solely on the environment. As mentioned below, the EBRD official involved in the Corridor 10 project felt that the standard of the EIA study completed by the Serbian technical consultants was quite high, so the focus was on bringing them 'to the next level'. For the Corridor 10 EIA studies, EBRD pressed the Serbian consultants on the sections on noise and biodiversity, which were improved eventually.[19] Thus, although the initial versions of the EIA documents were of a relatively high standard, EBRD undertook a tutorial role – as the World Bank did – to further improve the final version of the studies.

Officials from the IFIs were also involved in the public consultation for the Corridor 10 project. In addition to the EMP, the World Bank also required joining together the EIA studies from each sub-section in order to create a corridor-level study for which there would be an additional public consultation with a corresponding public meeting. A representative from the World Bank attended the corridor-level EIA meetings for the E-80 in Bela Palanka and for the E-75 in Vranje.[20] However, the involvement of the World Bank in the public meetings appeared limited to the corridor-level consultations required for its own procedures. According to an expert from one of the technical consultants (Highway Institute), the banks were present at the meetings in Vranje and Bela Palanka but did not attend the meetings for the sub-sections of E-75 and E-80.[21]

As in BiH, the IFIs took a dual role of stakeholder and tutor, although this latter role was with a lighter touch in Serbia. In particular, the documents that the IFIs received from the Serbian technical consultants were of a satisfactory standard, but nonetheless, both the World Bank and EBRD sought to improve the capacities of the ministry and technical consultants even further to consolidate longer-term impact in EIA practice. However, as seen below, the Serbian developer and technical consultants believed the rigour with which the World Bank procedure was followed 'to the letter' presented unnecessary obstacles instead of focusing on determining and mitigating the detrimental socio-environmental impacts of the road-building project.

Governmental structures: from technical to socio-technical

The EIA procedures for the various sections of Corridor 10 in Serbia provide a natural experiment to differentiate projects funded by IFIs versus those supported by the Serbian government, since the northern stretch of the road in Vojvodina was funded by the Serbian government, whilst the two sections in the south branching off from Niš were both financed through IFI investment. Thus, it is possible to see how the conduct of the governmental stakeholders differed in the two parts of the road.

In Vojvodina, the EIA process for the northern sections of the E-75 road were completed in 2006, but due to delays in funding, the project was not implemented before the two-year expiration, so it had to be repeated.[22] Officials from the Environmental Secretariat of Vojvodina were present for the three rounds of public consultations, the second phase of which had to be repeated. The consultation meetings for Corridor 10 were all held on the premises of the Executive Secretariat of Vojvodina in Novi Sad, and not in the affected municipalities. However, in accordance with the recent (post-2004) changes in EIA legislation, the Environmental Secretariat contacted the municipalities about the consultation, provided information about the meeting details in local newspapers, and sent copies of the draft EIA to the locales in question. The Secretariat was responsible for the organization of the meetings for the EIA process.

For Corridor 10 and EIA processes more generally, the Secretariat found that there was not sufficient interest amongst local governmental or community stakeholders to organize meetings in the affected municipalities.[23] However, in lieu of attendance at the public meetings (or interest in organizing meetings locally), the municipalities were engaged in the process, preferring to provide their opinions about the EIA in writing. The co-operation between the Environmental Secretariat and municipalities has improved since 2004, since the new legal framework required training and co-ordination.[24] The only governmental institution that attended any of the public meetings was the working unit of the Institute for Nature Protection of Serbia based in Novi Sad.

In contrast to the procedure in the autonomous province of Vojvodina, the EIA consultations on the southern sections of the E-75 and E-80 roads included more co-ordination between central governmental authority (in this case, federal authorities in Belgrade) and active participation by affected municipalities. As in the Vojvodina section of Corridor 10, the Institute for Nature Protection provided opinions and was involved to ensure that the impacts on nature had been considered. In the case of Corridor 10, the alignment of the motorway is mostly defined, so the Institute focused on ensuring that the developers gave due consideration to the environmental impacts of the project, and to suggest means of minimizing negative effects. In less strategic road-building projects, the Institute can propose alternative alignments that would be least detrimental to nature. For example, the Institute provided a preferred alignment for the route between Belgrade and the Adriatic coast road, but this was not the case for Corridor 10.[25] As part of the EIA process, the Institute of Nature Protection and the Institute of Cultural Heritage are required to be consulted. Thus, the Institute receives EIA documentation from the Ministry of Environmental Protection at the initial stages of the project. In the case of Corridor 10, the Institute found that there were no major problems regarding nature protection.[26] The Institute mainly liaised with the Ministry, not municipalities or technical consultants, since the latter are the responsibility of the Ministry. The Institute did provide opinions for the sub-sections of Corridor 10, but due to the number of sections (and thus, the number of public meetings), the Institute was unable to attend most of the hearings.[27] The role of the Institute of Nature Protection underlined two shortcomings in the context of multi-level socio-environmental processes. The first is that the remit of the Institute is quite narrow, focusing only on protection of nature and not environmental protection more generally.[28] Second, the Institute of Nature Protection offers its advice at the beginning of the EIA process directly to Ministry, and it does not have the opportunity to confirm that its input has been included in the final version of the study.[29]

The central co-ordination of the EIA process for Corridor 10 and for all other major projects in the Republic of Serbia is carried out by the Ministry for Environment and Spatial Planning.[30] As in the Bosnian case, the environmental protection procedures prior to the post-2000 legislation focused entirely on technical and procedural aspects of the process, and it is only with the more recent laws that the socio-environmental aspects of infrastructure projects have been included. Nonetheless, the existing technical capacities of the Ministry, coupled with proximity to research institutes in Belgrade and international exchanges, has left the Ministry in a strong position to complete the EIA processes in line with international standards of best practice. A representative from the Ministry chaired the Corridor 10 EIA public meetings and officials from the Ministry collected the comments during the hearing. The Ministry also took the lead liaising with the developer and the municipalities affected by the project, and also posted information about public meetings in newspapers and on television.

However, unlike FBiH, the EIA studies are not posted on the Ministry website.[31] However, all of the documents are available on the World Bank website[32] and the Corridor 10 LLC web page.[33]

The Ministry devolved the organization of the public meeting itself to the municipal authorities where the hearing took place. For the Corridor 10 EIA public meetings, the municipalities were in charge of organizing a venue that will accommodate local interested parties. Municipal authorities also contributed to sending information about the EIA hearings through local media.[34] However, municipal authorities had a dual role, both as organizers of the meeting and also as stakeholders within the EIA process that would provide opinions about the socio-environmental effects of the road-building project. Municipalities often worked with the technical consultants to provide local baseline data in order to aid the drafting of the EIA study.[35] Municipal authorities, not only from environmental departments, were contacted in writing about the EIA hearings and were expected to attend. Some of the municipalities involved in these EIAs had specific environmental departments, whilst others did not. Nonetheless, the situation had improved significantly in the last 5–10 years.[36] The actual municipal involvement in the EIA processes was mixed. Perhaps unsurprisingly, officials at the Public Enterprise – Roads of Serbia (which corresponds to Bosnian-entity Roads Directorates) identified the Southern Serbian cities of Vranje and Niš as municipalities which actively participated in the Corridor 10 EIAs.[37] Corridor 10 LLC, a public company which took over responsibility as the developer of the project from Roads of Serbia, also noted that Dimitrovgrad had good environmental officers in the municipal administration who cared about environmental issues.[38] Vranje and Dimitrovgrad are moderately sized cities in Southern Serbia, whilst Niš is the largest city in Southern Serbia and third-most populous city in the country (behind Belgrade and Novi Sad). Niš is also a university town in which the local municipality takes an active part in inter-municipality co-ordination on environmental and tourism issues, as mentioned in Chapter 4. However, active municipal participation is not limited to urban settings. For another road-building project EIA, Roads of Serbia noted that Crna Trava was very good, even though the Eastern Serbian town has approximately 1500 inhabitants and is amongst the poorest municipalities in the country (Beta, 2010). During the public consultations for the Corridor 10 road, the capacities of the municipal administration had been adequate even in Bela Palanka,[39] though the town is also amongst the poorest in Serbia. In general, the head of the section responsible for the EIA from Corridor 10 LLC noted that, irrespective of size, 'most local municipalities recognise their own obligations about environmental aspects of all projects on their territory including this [Corridor 10] highway project'.[40]

Public companies: developers for Corridor 10

Like their counterparts in the Ministry, the technical consultants within the firms completing the EIAs and the developers also benefitted from existing capacities in Serbia. According to an official at EBRD, one of the IFIs that invested in Corridor 10: 'Serbia has a tradition of highly educated people, people who worked in the private sector, even during the socialist regime and have worked overseas, so there is a lot of exposure to international culture'.[41]

This history of long-standing technical knowledge transfer from Western Europe to Serbia is exemplified by the head environmental specialist for the Corridor 10 LLC (K10), the developers for the E-75 and E-80 motorway projects. He trained as a civil engineer and became a specialist in EIAs, working at the Highway Institute (HIP) in Belgrade. As part of a programme funded by the German government, he was invited as one of five representatives from Serbia (in the topic of environment) to work in Germany for a year to further his technical knowledge, after which he returned to Serbia and started work as an Environmental Specialist at the Public Enterprise for Roads of Serbia. The exposure to work abroad has also helped him keep abreast of EIA practice in other places, and how lessons can be learned for completing EIA studies in Serbia.[42]

As in BiH, to comply with European best practice and to adhere to IFI standards with respect to EIA, the Public Enterprise for Roads in Serbia established an environmental department, with the now Head of Environment at K10 along with another environmental expert.[43] For these reasons, when the government established K10 to lead the project development of the motorway, it was necessary to include an environmental department to ensure that the documentation followed environmental and social guidelines set out by the World Bank and by EBRD.

Although the IFIs acknowledge that there are high levels of administrative and technical capacities in Serbia, legislation compliant with European best practice, clear delineation of bureaucratic responsibilities (unlike in BiH), and a long history of knowledge transfer, K10 (and the Serbian technical consultants) nonetheless had to navigate bureaucratic hurdles to complete the environmental assessment documentation to

138 *Europeanization of the Western Balkans*

fulfil the requirements of the IFIs. The developers at K10 highlighted two places where the environmental and social assessment for the E-75 and E-80 projects had been negatively affected by the rigidity of IFI procedures: the additional consultation process for Environmental Management Plan (EMP) documents for each sub-section of the road, and the consultation for two corridor-level EIA documents (one each for E-75 and E-80).

Officials at K10 felt that much of the public consultation and preparation of documents were duplicated during the EIA processes for the highway, since the IFIs needed to have 'every box ticked' before they could provide the loans for the highway project.[44] According to the developers of the highway project, the IFIs did not treat the domestic legislation as sufficient for the environmental and social assessments, instead requiring their own procedures. From the standpoint of those who had completed the EIA, the differences between the domestic and IFI procedures were not significant, yet the resulting additional work was both time-consuming and costly. According to K10, the World Bank requires EMP for the environmental impact study, particularly since a highway is a major 'Category A' project. The EMP consists of an outline of the mitigation, monitoring, and institutional strategies in the implementation of a project to minimize detrimental environmental and social impacts.[45] In the case of the projects to finance the southern sections of the Corridor 10 highway, K10 argued that the required information was already contained in the EIA study and that the technical consultants were working to the domestic legislation (which does not require an EMP). However, to receive the funds, K10 had to relent and organize additional public consultations for EMP documents for each section of the road with IFI investment. This added significantly to the time needed to complete the public consultation and the environmental assessment. Moreover, the costs for the extra procedure (and consultation) are borne by the developers. The K10 developers conceded that the EMP is an 'elegant' way to summarize the requirements that would form parts of the tendering process, but found it unnecessary to duplicate the consultation process when multiple rounds of hearings and deliberation had been completed for the EIA.

In addition to the time and resources K10 felt had been wasted on the consultation of EMP documents for each sub-section of the IFI-funded Corridor 10 highway, the Corridor-level EIA (along with another public consultation) was also seen as superfluous. As with the EMP, K10 felt that the EIA procedure had fulfilled the socio-environmental requirements dictated by European best practice, but not the procedures of the

IFIs. The three IFIs (EBRD, EIB, and the World Bank) had agreed that the project developers would also need to produce a Corridor-level EIA by bringing together all of the environmental concerns from each sub-section from the preliminary into one document. Since such a document would cover the whole length of the E-80 and E-75 projects, it could not be very detailed. According to the developers, this additional requirement had no function for environmental assessment, but rather was a procedural one for IFI auditing. This is because it is more useful to look at the sub-section documents, where it is possible to include greater detail about socio-economic conditions along the proposed route of the highway. However, despite the misgivings towards the additional procedures, officials at K10 realized that they were not in a strong position and complied with the imposed conditions: 'I recognise that it is a part of the procedure, and that is why we do not have room to argue or discuss. You are supposed to do it. It takes a lot of time before the approval. [The] actual benefits of those differences, I still cannot see them'.[46]

Technical consultants

For major road-building projects in Serbia such as the planning and construction for the Corridor 10 motorway, there are three main institutions that have the capacities to undertake and complete the required assessment studies: Highway Institute, Belgrade (HIP); Traffic Institute, Belgrade (CIP); and the Centre for Roads of Vojvodina, Novi Sad (CPV). The two firms in Belgrade were responsible for the EIA studies along the IFI-funded branches of the motorway (E-75 and E-80) in Southern Serbia, whilst CPV oversaw the EIA for the domestically funded northern part of the road between Novi Sad and the Hungarian border. The technical consultants had difficulties fulfilling the requirements for the assessment process in different ways.

Along the northern span of the motorway project, the planning and construction were governed by the domestic legislation, which, as mentioned above, has been largely harmonized with European standards since 2004, particularly in the procedure for public consultation. At CPV, the environmental department had three staff members who handled the procedure, which was unchanged from the number of team members before 2004.[47] This meant that the workload for each assessment has increased without a sufficient increase in personnel. At the time of the interview at CPV in late 2009, only the lead engineer had the qualifications to be able to sign off on assessment studies. This lead engineer noted that the government had rushed through legislation to adhere to international standards, without providing adequate means for technical consultants to acquire the relevant knowledge to be able to implement the new EIA procedure. For example, the noise and vibration requirements were transposed from existing German standards, which are quite stringent, instead of ratifying legislation that is more appropriate for local circumstances.[48]

However, notwithstanding the difficulties CPV faced in completing the EIA procedure, Serbia has a legacy of high levels of technical capacity in science and engineering, which stretches back to the socialist period. For example, HIP (in its present form) started in 1972 and has hosted a number of international symposia and professional conferences related to road-building.[49] Although its origins are older, the Yugoslav state established CIP as a scientific research centre in the 1960s, merging with other institutions in the 1980s.[50] According to EBRD, one of the IFI investors for the E-75 and E-80 motorway projects, the state and technical consultants in Serbia are 'quite good' at mobilizing the documentation to complete robust EIA studies, and furthermore: 'There are consultants in Serbia that can do EIAs. There are even consultants in Serbia that can...look at the social aspects which are now getting more and more prominent in our policy'.[51] From the standpoint of the IFIs, Serbia already has a high level of technical capacity (which is the legacy of the focus on science and engineering during the socialist period), but the EIA procedures carried out by domestic actors may need fine-tuning: 'We still try to take them to the next level...So for instance, when we work with a Serbian client, we look at things like certification, international standards ISO 14000, or Safety 18000...We still try to get them to a better standard'.[52]

The two major technical consultants in Belgrade (HIP and CIP) have significant experience in putting together EIAs for road-building projects. For example, HIP started producing EIA studies for highway-building projects in the 1990s. At that time, assessments were carried out at two stages: to determine the route of the proposed road, and the preliminary design. HIP's first two EIAs were the route location studies for the Belgrade Bypass in 1992 and the Niš-Macedonian border road in 1994. At that time, EIAs were only a technical exercise, since there was no requirement to carry out a formal consultation as in the current procedure. The only opinion that technical consultants included in every EIA was that of the Institute for Cultural Heritage, a government-funded research institute. The Institute for Nature Protection would only be approached if the proposed route passed through a national park. In the period between 1992 and 1996, HIP had four members of staff responsible for EIAs, and the number of staff members was eight at the start of 2010, though there was no expansion in response to the new legislation in 2004.

HIP led the EIA procedure for four of the five sub-sections of the road-building project between Niš and the Bulgarian border (E-80).[53] They made presentations at the public meetings held in various locations along the proposed route of the road to explain the possible environmental impacts to potentially affected populations. The environmental department at HIP generally had a positive experience working with the experts at the IFIs. According to a consultant at HIP, a representative from one of the IFIs came to help with the EIA study for one of the sub-sections of the road. By its own admission, the HIP 'document [draft EIA study] was not so good'. The expert from the IFI made a number of suggestions to improve the EIA study, thus underlining the tutorial role the investors played in completing the Corridor 10 environmental assessments.[54] However, as mentioned in the previous section, the IFIs (using the World Bank policy) added the extra stipulation of completing EMP documents for each section of the road. The EMP is generated using the detailed design of the road project, whilst the EIA is created from the preliminary design. The main detailed design of the road consists of more than the five sub-sections used for the preliminary design, and for each sub-section of the detailed design, there is an EMP public consultation in addition to the EIA consultations. To comply with the EMP requirements, HIP produced an Environmental Protection of the Project document within the detailed design, which has all of the information necessary to generate the EMP.[55] Thus, the interaction between the IFIs and technical consultants was twofold. On the one hand, experts from the IFIs provided a strong tutorial role in improving draft documentation, and this learning would be beneficial for future studies taken on by HIP. On the other hand, IFIs required additional steps that technical consultants needed to complete in order to fulfil the requirements for the international loans. Although it was not stated as explicitly by the HIP as K10 had done, producing the EMP more than doubled the number of documents requiring public consultation, though K10 sought to organize EMP and EIA public meetings at the same time and venue whenever possible.

NGOs and local communities

For the most part, local and national Serbian NGOs did not play an active role in the EIA public consultations for the Corridor 10 motorway.

One of the technical consultants from HIP, who presented the findings of the EIA study, noted that NGOs may have been present at public meetings, but that there had been no contact between the technical consultants and the NGOs.[56] The IFIs echoed the conclusions of the technical consultants. NGOs were present at the public meetings and are listed in the studies. The NGOs sent their comments directly to the developer of the project (i.e., Roads of Serbia and later K10). However, the role of NGOs during the EIA public consultation was 'not prominent' in any way, and these organizations tended to not be very vocal. In total, the EIA study contains approximately five comments from NGOs.[57] An EBRD official also concluded that ENGOs were not active in either BiH or in Serbia compared to other post-communist countries in Europe. The official conjectured that this might be the result of substantial funding for NGO development in other transition countries by international organizations (e.g., UN and the Open Society), which has led to NGOs (particularly human rights organizations) acting as a strong 'civil society' counterweight to the governments in places such as Russia and Georgia. By contrast, despite the two decades of donor-driven civil society development in the Western Balkans, there has nonetheless been a lower level of commitment in the region to create a sector of strong, well-structured NGOs. The IFI experience is that there are Serbian NGOs that developed in some sectors – particularly human rights and Roma protection – but environmental organizations continue to be relatively weak.[58]

The one notable exception to the lack of NGO involvement in the Corridor 10 EIA procedures (including the Belgrade Bypass) is CEKOR. As mentioned in Chapter 4, CEKOR is based in Novi Sad but has worked co-operatively with other active ENGOs in the region around common campaigns such as the alignment of the Blagaj-Počitelj road in BiH (alongside CZZS, Ekotim, and Zelena akcija). CEKOR is also linked with pan-European environmental networks as a member institution in CEE Bankwatch, which oversees IFI investment in post-socialist Europe. In the case of Corridor 10, CEKOR has been vocal in its misgivings about the EIA procedures for the Belgrade Bypass and Gazela Bridge sections of the route, with a particular focus on the additional socio-environmental impacts associated with the resettlement of Roma who had been living in the vicinity of the bridge. CEKOR wrote an issue paper in 2007 summarizing its critique of the EIA process for the aforementioned sections of Corridor 10 (CEKOR, 2007). According to CEKOR, the EIA completed for the Belgrade Bypass has the following environmental problems: lack of research on the agricultural impacts (Section A); insufficient measures to protect Belgrade drinking-water supplies during construction (Section B); and there has not been enough of an assessment on the environmental impact on cultural/archaeological sites (Section C). The Gazela Bridge was not included in the EIA study at all, since, according to the EIB, it was not required under either Annex I or Annex II projects to carry out such an assessment for the upgrade or rehabilitation of a bridge. However, CEKOR countered that there would be increased air pollution with ramifications for public health for residents living near the bridge. CEKOR recommended that a full EIA and SEA be completed for the whole route (including Gazela Bridge), but this was never completed (CEKOR, 2008).

Although it is not a purely environmental issue, it is instructive to examine the role of CEKOR (as part of CEE Bankwatch) in the Gazela Bridge resettlement issue. The plans for the rehabilitation of Gazela Bridge required moving approximately 200 Roma families living in slums under the original bridge. To facilitate the resettlement of these families in line with IFI standards, EBRD provided technical assistance to the City of Belgrade to produce a Resettlement Action Plan (RAP). In response to CEKOR's queries, EIB clarified that the banks had pursued a 'harmonized' approach to ensure to fulfil requirements related to social standards for involuntary resettlement in order to receive the loan.[59] Despite these assurances, the families were forcibly evicted in August 2009 and rehoused in metal containers without adequate standards of hygiene and other amenities, including education (Amnesty International, 2010b). Part of the problem could be ascribed to unclear and conflicting jurisdiction between city and state agencies involved responsible for aspects of Corridor 10, which may not be indicative of governance in the country. The state-owned Roads of Serbia (and later K10 LLC) led the development of the project, whilst land use and housing is the domain of city authorities. Without a clear plan of co-ordination, problems with implementation were bound to surface.

CEKOR lodged an official complaint with EIB in September 2009. In the meantime, the issue of the involuntary resettlement became a cause for the international human rights community and became part of Amnesty International 'Demand Dignity' campaign with its publication of a report in June 2010 calling for Serbian authorities to stop 'forced evictions' of informal Roma settlements (i.e., those built without planning permission) (Amnesty International, 2010a). Thus, the local issue about moving informal Roma communities in order to complete IFI-funded road-building projects from a domestic to an international issue linked with European and international human rights. EIB eventually responded in July 2010 to the CEKOR complaint, conceding that there was non-compliance with the policies of the World Bank, and proposed to audit resettlement and 'livelihood restoration' through on-site visits, working with governmental authorities and affected communities. Further funding would only be disbursed when EIB was satisfied that resettlement had proceeded in line with the World Bank's standards (EIB, 2010). Although this signalled a victory for the CEKOR campaign (through its links with pan-European and international NGOs), further involuntary resettlement of informal Roma communities – as part of the EBRD- and EIB-funded Sava Bridge project – continued to take place (CEE Bankwatch, 2012).

There is mixed evidence regarding the levels of active citizen involvement in consultative processes such as the Corridor 10 EIA. On the one hand, there is a legacy of drawing attention to grievances at the local level in Serbia. There is an existing law on Public Access to Information that pre-dates Serbian ratification of the Aarhus Convention, so there will probably be little change in the conduct of local citizens in the context of public consultations. According to an official at the World Bank, Serbian citizens are aware of the law and are 'not afraid to write letters' to governmental institutions to air their views.[60] On the other hand, EBRD found the local population to be compliant during the EIA processes in the country. In particular, citizens seem to have a certain level of trust in the government's work on large projects and expect a certain amount of personal sacrifice in order to complete the road construction (e.g., resettlement).[61]

As with the case studies in BiH, and unlike road-building projects in CEE and Western Europe, local citizens in Serbia tended to eschew a 'not-in-my-backyard' (NIMBY) attitude and favour the alignment of the road to be as close as possible to their homes. This is because economic concerns trump environmental ones for most residents along the proposed route. Southern Serbia is the most impoverished region in the country, and local communities especially see the section of the Corridor 10 road to the Bulgarian border as an economic 'lifeline'.[62] This local support is further galvanized by the aforementioned trust in government infrastructure projects, with many believing that the road is being built for 'the public good'.[63]

At the public meetings along the IFI-funded sections of Corridor 10, local citizens did make some positive contributions to the study that enhanced the final draft of the EIA. There were often qualified members of the public present at the hearings, like researchers from university faculties. The attendance at such hearings varied, but there were often more than 10–15 attendees.[64] The actual relevance of the questions varied and did not always focus on environmental aspects of the project. The technical consultant for the EIA noted: 'Some of them recognise what the environmental part of the project is. Some of them are not even close. They just come to see whether it is related to their own problems'.[65] In particular, local residents along the proposed route were interested in expropriation (i.e., how much compensation they would receive if the road or adjoining exclusion went through their property). These property issues caused problems between the developer and local citizens at the public hearings.[66] Although resettlement and expropriation are not part of the EIA study (and are part of separate documentation), local residents often asked repeated questions about property issues and were not interested in other issues.[67] However, the level of interest in the EIA public consultations in Southern Serbia contrasted significantly with the meetings in Vojvodina. The only attendees for the meetings in Novi Sad were the Secretariat, the developers, and the Institute for Nature Protection (working unit in Novi Sad). There were no members of the public or NGOs present. The officials at the Environmental Secretariat in Novi Sad noted that local citizens never attend meetings for larger projects, but they are more involved in smaller installations such as mobile phone masts. If there had been sufficient interest in the local communities through which the road would pass, the meetings would have been organized there.[68] There are two possible reasons for the differences in the level of public engagement. The first is that since the IFI-funded portions of the road had international observers at each meeting, local citizens may have felt more empowered to attend meetings and challenge the findings of the study (though the grievances were often outside the scope of the EIA procedure). Second, as the World Bank official mentioned above, the road is seen as a 'lifeline' in many parts of economically deprived Southern Serbia, whilst Vojvodina is relatively affluent, so the road would be beneficial without being as fundamentally crucial to residents compared to the southern sections of the road.

As mentioned above, the public consultations along the proposed Corridor 10 route (E-75 and E-80 roads) have already been completed,[69] so it was not possible to collect the same primary focus group data as in BiH. However, the reporting requirements from IFIs (namely the World Bank) have necessitated thorough documentation and transparency. To this end, drafts of the environmental assessment studies (i.e., EIA and EMP) are available on the Corridors Serbia website, and EMPs are available both on this and on the World Bank web pages. Although this information is not always detailed, it does confirm some of the observations from environmental assessment public consultation in Serbia more generally. There were approximately 10–20 attendees at each of the public meetings. A World Bank representative, the technical consultants of the study, and one or more representatives from Corridors of Serbia (formerly K10) were also present. In the meeting in Vranje in March 2009, of the 13 participants, there were only three individuals attending as private citizens, who were joined by representatives from Vranje and Vladičin Han municipalities, and a local motel.[70] By contrast, the public meeting in Grdelica in May 2009 was attended by 20 individuals, of whom 15 attended in the capacity of private citizens.[71] In a more recent EMP public consultation in April 2014 in Vladičin Han, eight of the 13 attendees were present in their capacity as private citizens.[72] Of the documents that included the list of participants, none mentioned the presence of NGOs, and the explicit input of civil society organizations is absent from the available public consultation documents.

Some of the concerns raised by citizens were relevant to environmental impact. Borrow pits are locations where materials have been excavated for use in other locations. For example, gravel might be excavated from a location to be used in the road construction for Corridor 10. Attendees at the meetings in Vladičin Han and in Bela Palanka asked about the locations of these pits. Participants at the meeting in Vladičin Han also raised concerns about noise and surface water flooding that might be worsened by the construction of the motorway. The issue of water loss in wells in Kržince was also raised, but it referred to a neighbouring LOT outside the scope of the discussion, which may corroborate the concerns of some ENGOs that developers employ 'salami tactics' to exclude more contentious sections of road-building assessment processes. An additional concern was raised by a municipal official at the meeting in Bela Palanka about ensuring the protection of the spring Stulanovac in Crvena Reka. Local citizens at the meeting in Vranje aired a similar worry about potential damage to water sources due to the road construction.[73]

Although there were some comments about environmental issues, there was a strong focus by citizens on the social and economic impacts of the alignment of the road. For example, as part of the consultation for the section of E-80 between Prosek and the Bulgarian border, the catering firm 'Happy Star' sent in a written comment urging a change of alignment, so that its premises would not need to be demolished. The firm used an economic rationale, stating that there would be a significant detrimental social impact, since there would be the loss of 20 jobs in this deprived part of Serbia.[74] Municipal representatives and citizens from Vladičin Han also framed their concern that the road alignment between Caričina Dolin and Momin Kamen would be away from the settlement, leading to development elsewhere, thus resulting in detrimental economic effects and outward migration.[75] Local citizens involved in consultations also queried the location of the motorway as reconfiguring access to their private agricultural lands.[76] As a way to address this problem, citizens in Vladičin Han requested alternative road construction to ensure local traffic links remain in place.[77] As with all road-building projects both in Serbia and in BiH, land acquisition and compensation was raised often in these consultations, even though it is not an environmental issue,[78] including land acquisition for borrow pits.[79]

Conclusion

Thus, the road-building EIA processes around the Corridor 10 motorway in Serbia share a number of characteristics with the processes in BiH studied in the previous chapters. First, for the most part, the legislative framework for environmental assessment and public consultation adhere with EU directives and best practice. Second, local governmental officials and technical consultants both indicated that the rapidity of reform in the assessment processes diverged from the realities on the ground, both with respect to administrative capacities and the level of participatory engagement of non-state actors. The EC Progress Reports continue to highlight the need to improve public consultation during EIAs. Third, the process of Europeanization (through SAp) in both countries has focussed, quite unsurprisingly, on issues that were the consequence of the armed conflicts in the 1990s, namely the arrest and extradition of high-profile war-crimes suspects to the ICTY, refugee return, and normalization of relations between Western Balkans governments. Finally, and perhaps most importantly, the prevailing EU instrument of soft conditionality for potential candidates to create robust multi-level environmental governance does not trigger transformation on its own.

The EIA procedures in Serbia (and in BiH) worked best and laid the foundations for changes in long-term practices when the projects were funded by IFIs that took an active tutorial role, including presence at the public consultations. However, local technical consultants complained that some of the steps of the auditing requirements for the World Bank, particularly the drafting of EMP documents, did not add anything to the process, but rather added unnecessary and time-consuming extra conditionality in order to 'tick boxes' for IFIs' own procedures. According to one of the IFIs, technical consultants and governmental officials from Serbia already produced good EIA documentation, but it was thus instructive for IFIs to assist in taking the environmental assessments 'to the next level'.

It is here that the transformation of EIAs, and governance more generally, differs between Serbia and BiH. In the former, there is a long tradition of high levels of capacity in engineering. Thus, EU instruments should be designed to make adjustments (however significant) to existing stakeholder institutions in order to align them with best practice in the rest of Europe, rather than building new institutions. For example, both HIP and CPV had environmental departments at the time of the new EIA legislation in 2004, so the new laws brought about new procedures for these technical consultants, not new institutionalization. By contrast, as mentioned in Chapter 3, the FBiH Roads Directorate established an Environmental Department in response to recommendations from the World Bank.

Although ENGOs were also largely missing from the EIA public consultations around Corridor 10, the situation differs significantly from the case studies examined in BiH. Unlike the other ex-Yugoslav republics, with the exception of Slovenia (Figa, 1997), there is an activist heritage in Serbia that resonates with experiences in other parts of SEE and CEE, though the impact of these movements were attenuated by nationalist regime in the 1990s. Still, organizations like MIS and Green Network of Vojvodina grew out of activist networks during the socialist period, and Ecotopia (despite its links with established party politics in present-day Serbia) was started by leaders of Otpor!, the movement that led the October 2000 overthrow of Milošević.

With the increased focus of EU instruments during candidacy and hard conditionality to comply with the *acquis communautaire*, the next few years will be marked with rapid changes in the institutionalization of environmental governance. Closing chapter 27 (Environment) of the *acquis* will be particularly difficult. The experience from Croatia, the most recent state to join the EU and the first Western Balkans country to do so, is that the negotiation process allows a limited opportunity for CSOs to be involved in shaping the final outcome (Đokić & Sumpor, 2013). Moreover, civil society used the accession process in CEE and SEE to leverage the government in environmental issues (see the contributions in Börzel, 2009a), though they have often been too weak to do so (Börzel & Buzogány, 2010a). Whether the EU negotiations around chapter 27 will lead to a robust multi-level environmental governance remains to be seen.

Conclusion

The research underpinning this study set out to undertake empirical and comparative analysis of a particular policy area – environmental protection and regulation – in order to interrogate the impact of Europeanization on BiH and Serbia. The aim was to identify the distribution of power amongst the various domestic and exogenous governance actors, and to decipher the drivers of change and the various veto players. The focus throughout has been on understanding how power is exercised formally and informally, why change occurs and why it does not, and which interventions are critical in instigating a shift towards new modes of governance. The starting premise, or hypothesis, was that post-conflict, post-authoritarian successor states of the former Yugoslavia represent a difficult test case for liberal political and economic reform, and for the transposition of Western models of multi-level governance. These states also pose a profound challenge to the strategies and instruments used by the EU as part of its Eastern enlargement. At the time of conducting the research, the progress of neither country was directly dependent upon progress towards environmental governance reform. Subsequently, Serbia has become a candidate state and is therefore now under direct pressure to ensure compliance with all chapters of the *acquis*. Bosnia's progress towards full candidacy remains stalled and the failure to achieve substantive reform in the realm of, for instance, environmental governance is indicative of the fundamental obstacles the country faces in its engagement with the EU, as well as in dealing with domestic political and economic realities.

It is therefore necessary to draw three sets of conclusions. First, a set of conclusions based on the empirical data concerning EIA processes in Serbia and BiH and the development of environmental governance. Second, as the main driver of change, is the EU and the process of

Europeanization. We need to provide commentary and draw conclusions about both the impact of Europeanization and the process itself. Finally, there is a need to relate the specific findings to the literature and commentary on BiH and Serbia: what can we learn, if anything, about the political development of these countries from our perspective of environmental governance? As with any scientific research study, there is a need also to highlight both the limitations of the current findings, but also to indicate how these findings and our approach may be built upon and extended in the future. This concluding chapter will aim to address each of these tasks.

EIAs and environmental governance in BiH and Serbia

What characterizes the emerging forms of environmental governance in Bosnia and Serbia? The first point to note is that in both locations environmental governance is framed in terms of new legal instruments that are basically compliant with EU directives. This is not to suggest that further legal reform is unnecessary; rather, that there is a momentum to harmonize and to enact in a policy realm that is not without its contentions, is resource intensive, and where there is a considerable gap between what these countries had in their legal codes to start with and what has needed to be adopted. European-wide standards of best practice are incorporated into new legislation, and in some cases, even exceeded.

However, formal legal harmonization with EU standards and best practice should not detract from the requirement for reform of processes and institutional capacities so as to ensure that effective implementation occurs. Whilst new legal frameworks may well be in place, the implicit requirement of such new laws to engage stakeholders in a deliberative process, develop new institutions, or change the distribution of resources between the various actors and interests has invariably not occurred. There is clearly a shift towards opening up decision-making to stakeholder networks and incorporating new sources of domestic and external knowledge within the governance process, but this remains in its infancy and is by no means routine or universal. From the perspective of EIAs, it appeared that any attempt to shift towards new modes of governance was attenuated by the lack of administrative capacities and other impediments at the domestic level, including the absence of appropriate knowledge and expertise, or the lack of political will. Relatively closed and centralized hierarchies in both countries seemingly dominate environmental governance, with little regular or significant involvement from NGOs and other societal networks.

In BiH, the alignment of capacities, authority, and responsibility is poor. This is never simply a rational or technical exercise, and effective devolution of responsibilities and the transfer of authority to sub-state or non-state actors can only take place within a constitutional framework and culture that permits this to happen. The prerequisite of such new governance is the existence of a shadow of hierarchy and a notion of how decisions will ultimately be enforced if the informal or non-institutionalized processes fail. What appears, at first glance at least, to be a multi-level system of environmental governance in BiH is a mirage: what exists in reality is duplication, weak authority, misplaced and low levels of capacity, and the absence of a shadow of hierarchy able to enforce state-wide standards.

Indeed, what our research illustrates very clearly is how the complexities of the Bosnian constitution have stymied an effective transfer of competencies, capacities, and authority. Environmental regulation and enforcement occurs within a poorly co-ordinated labyrinth of centralized hierarchies – locales of centralization – amidst which various capacities and resources have been scattered, mostly by chance rather than by design. The situation in Serbia is far better and there is a clear institutional framework for environmental governance that maps onto a rather centralized state structure that is embarking upon a process of decentralization (local administration reform and the empowerment of municipalities is an ongoing process). Where there is resistance to devolution and new modes of horizontal power in Serbia it is more a consequence of the legacy of state socialism than war and the breakup of Yugoslavia.

Although already referred to above, the weakness of NGOs needs further discussion here. The handful of environmental organizations in both Serbia and BiH that attempt to influence policy and to engage in politics possess very low levels of advocacy and scientific expertise. This has been attributed here to the absence of a green dimension to the end of socialist Yugoslavia, and to the failure of pan-European green networks to operate in the region. But it is also undoubtedly a reflection of the limitations of international donor strategies, which have ostensibly (and at times ostentatiously) targeted 'civil society' and NGO developments as priorities. In the case of environmental NGOs in CEE and SEE, it was possible to point to a set of project management skills as legacy to all the project grants and funds channelled through 'civil society' by foreign donors. From the perspective of environmental NGOs in BiH and Serbia, there was little evidence even of this.

What the research also reveals about environmental governance in these countries is the extent to which exogenous actors play a critical role, whether through funding and training, or because they are the main source of finance for the infrastructure projects that require EIAs. It is thus international consultants, IFIs, banks, and foreign companies that are the main stakeholders alongside domestic governments and citizens. Such a reality has positive and negative consequences. It ensures that international, compliant standards are adhered to and made to work as a consequence of EBRD or World Bank involvement. One negative aspect is that EIA processes and the requirement for deliberation are viewed as yet another expression of external interference and involvement in domestic processes. Although we found little overt evidence of this, there was certainly a sense that all international actors – investors, consultants, construction companies – were seen as part and parcel of a less than transparent liberalization process.

Europeanization and the impact of the EU

The EIA processes studied as part of this research occurred when both countries were 'potential candidates' for EU membership and therefore subject only to soft conditionality. As potential candidates, BiH and Serbia were not being pressured to realize rapid legislative reform, but were nevertheless assessed as part of the annual progress reports according to the progress they had made against each chapter of the *acquis*. The key difference between being a potential candidate and a candidate is that the former status does not provide access to full IPA funding, and negotiations for EU membership can only begin once full candidacy is achieved. In other words, as potential candidates, no amount of progress will lead directly to membership negotiations. There is arguably, therefore, a reduced incentive for progress towards compliance. Serbia's status changed in 2011 with the capture of war-crimes suspects Ratko Mladić and Goran Hadžić, and with the landmark Brussels Agreement in April 2013 that started the path of normalization between Belgrade and Pristina. The former paved the way for candidate status in March 2012, and the latter led to the start of accession negotiations in January 2014.

What do we learn, or is inferred, from our case studies about Europeanization and the EU's transformative power over third-party states? Whilst the EU undoubtedly exerts considerable influence over laws, processes, and perceptions of the 'way to do politics', its actual influence on the behaviour of officials, citizens, experts, and economic actors is heavily contingent on the involvement of other external agencies and organizations. In this sense, the EU sets a framework; but in terms of implementation and social learning, its reach is limited and its assistance too remote to singularly generate substantive change. From the perspective of our findings vis-à-vis EIAs, we can only conclude that domestic officials and actors need to be shown how to make EU-compliant processes work in the first instance. The knowledge and expertise needs to be applied and, particularly where there is significant resource deficiency, processes need to be tried and tested under the auspices of a more experienced external partner, be that an international NGO network, the EBRD, or a French consultant.

As already noted, the EU is reasonably effective at building administrative capacities and providing training and resources, but its ability to drive the necessary political or constitutional change to ensure that these resources and capacities are located amongst progressive and pro-change elites is, not surprisingly, limited. This is largely because the enlargement strategy and the tools and strategies that the EU deploys in BiH and Serbia were fashioned to support the consolidation of democratization in CEE, where pro-change elites were in positions of power, broadly speaking, and were engaged in a process of bureaucratic reform and liberalization. The Stabilisation and Association process (SAp) and IPA funds were not engineered to deliver regime change in BiH and they are not suited for such a task as a consequence. Our empirical perspective illustrates this harsh reality very clearly: the EU has disseminated some assistance, it has directly or indirectly built the capacities of state agencies, but it has not been able to engender the necessary constitutional reform to ensure that environmental regulation functions effectively across the whole territory of the country.

This recent progress by Serbia in the European integration process, whilst encouraging on one hand, also underlines the weakness of current EU instruments as drivers of fundamental *societal* transformation. The primary mechanism used by the EU is the SAp, which consists of the SAA, free-trade agreement, IPA assistance, and regional co-operation. Despite the many facets of the SAp, it is the prioritization of regional stability and security that has dominated the trajectory of EU integration in the former Yugoslav territories in the Western Balkans. This is not surprising, given the long shadow cast by the conflicts in the 1990s and early 2000s. In Serbia, extradition of the three fugitive war-crimes suspects to the ICTY and normalization of relations with Kosovo allowed the EU integration process to continue, rather than provide evidence of substantive progress towards establishing new modes of governance. In BiH, the signing of the SAA was repeatedly stymied by intransigence of RS leaders regarding changes that would create a unified police force. In both places, security sector reform is a central part of the reform process.

Both states joined the Partnership for Peace along with Montenegro in December 2006, which entails closer relations with NATO, and which is a stepping stone for full NATO membership.[1] Thus, the primary focus of the SAp has been that of regional stability and security over other issues, unlike the European Partnership agreements that preceded accession in CEE. The effect has been to sideline fundamental issues in the transformation of Western Balkan societies, such as minority rights as well as environmental governance and sustainability.

Moreover, using the toolkit of conditionality deployed during the previous Eastern enlargement of the EU is unlikely to deliver the same progress in the Western Balkans. As argued throughout this book, the transition from authoritarianism in former Yugoslavia during its disintegration occurred differently from its neighbours in SEE and CEE. In other parts of post-socialist Europe, the relatively peaceful transition to post-socialist, multi-party representative democracy came about through the co-operation between pro-reform political elites, social movement activists, and the general population. In the Czech Republic and Bulgaria, for example, environmental organizations were 'vehicles of transition', and environmental grievances were central to dissident coalitions during the collapse of the Soviet-style authoritarian regimes (Baker, 1998; Fagan & Jehlicka, 2003). By contrast, in Serbia, BiH, and other parts of former Yugoslavia, multi-party politics, the violent conflicts, and the terms of the peace agreements that followed were controlled by nationalist elites, with left-liberal civil society organizations excluded from formal politics, but also disconnected from society at large. Thus, as our research has shown, the EU's task of driving forth new modes of governance so as to involve NGOs (environmental and otherwise) requires a much more fundamental empowerment of weak, poorly networked activists. Rather than helping to propel relatively capable and established environmental NGOs towards policy elites and into new epistemic communities, the task in the Western Balkans is to actually create the organizations and networks from scratch, to then encourage experts to engage with the NGO coalitions, and to coax citizens out of their personal and enmeshed networks (Howard, 2003). In other words, setting conditions and providing incentives to encourage greater involvement on behalf of NGOs is not enough; what is needed is a stronger and more innovative 'bottom-up' approach designed to address the fact that pro-reform networks within both the governmental and non-governmental sectors are not only weak in terms of their capacities, but cannot easily attach themselves to or connect with party and political elites. The typical trajectory of environmental NGOs working with centre-left and liberal progressive parties is stymied by the continued dominance of (albeit relatively liberal) nationalist parties in both countries. This is especially true in BiH, where ethno-territorial fragmentation at the end of the wars in the 1990s has impeded the development of cross-ethnic ENGO networks within the country and across its borders. Moreover, the proliferation of ethno-national majority fiefdoms – what Kostovicova and Bojcic-Dzelilovic (2011) refer to as 'transnational networks' – has weakened the power of the state and also hampered the ability of relevant governmental officials responsible for environmental regulation to work together so as to cast an effective 'shadow of hierarchy' over decision-making and enforcement.

Despite the emphasis placed on the involvement of civil society and non-governmental organizations, the prevailing strategy of the EU of creating compliant processes for multi-level governance, actually consolidates the power and influence of existing elites. Following the template from the fifth enlargement, the SAp places primary emphasis on formal compliance, legal change, and institution building as a precursor for the emergence of 'good' or 'new' governance. However, in a situation where there has been disrupted or limited regime change, where 'old' elites remain in positions of power within the state and political machinery, an emphasis on institutional consolidation and legal compliance serve actually to bolster the status quo at the expense of progressive actors with civil society or indeed amid the political elite (Bechev & Andreev, 2005; Bojcic-Dzelilovic & Kostovicova, 2013; Juncos, 2012). Moreover, by depicting policy compliance and institutional change as predominantly technical or bureaucratic rather than political, the EU inadvertently weakens the agency of progressive actors and networks and reinforces their extrication from the policy and political process.

This research on EIA processes and environmental governance reinforces the point that the EU must frame its pursuit of technocratic standards and 'best practice' in terms of political and democratic reform. Despite the policy and administrative overtones of 'good governance', reforming decision-making and enforcement in the Western Balkans is essentially about completing disrupted and derailed regime change; it involves ensuring that its tutelage and assistance targets progressive actors and networks that are in, or close to, positions of power and the institutions of government. The focus on top-down efficiency and formal compliance allows for the development of what can be termed 'negative capacity' whereby those in positions of power within ministries or other governmental institutions are endowed with resources to secure their status within the policy-making hierarchy, and thus weakening the agency of, in this case, environmental NGOs.

Thus, one of the key conclusions of the research is that conditionality alone cannot deliver new modes of environmental governance in line with EU best practice. This in itself is not a particularly radical conclusion to draw, and is certainly consistent with what is argued within the existing literature (cf. Börzel and Buzogány, 2010b). By design, EU conditionality is deployed as a mechanism to augment the capacities and empowerment of government officials, based on the assumption that those within the governing hierarchy are sufficiently progressive in their thinking and approach to engage NGOs and the reform agenda. Whilst EU assistance for the environmental sector in BiH and Serbia has built the advocacy and service provision capacities of a limited number of environmental NGOs, there has been little progress towards new modes of governance. What this research has highlighted in particular is the value of direct and hands-on tutelage (or simply 'teaching') provided by international actors, including investors and financial institutions. That such assistance is of value in bringing about changes in environmental governance and the behaviour of state and non-state actors was also evidenced in the CEE countries. An important form of tutelage for NGOs during Eastern enlargement in CEE countries was given by transnational European ENGO networks, which were part of 'Green 10'. The exchange of know-how and modelling of good practice directly and 'on the ground' helped to inculcate new approaches to advocacy and mobilization of resources (Bomberg, 2002). It is therefore significant that we found that this form of social learning has been virtually non-existent in both Serbia and BiH for what appear to be two main reasons. First, although many of the Green 10 organizations worked with CEE partners during the first wave of Eastern enlargement, by the time the EU integration process reached the Western Balkans, these pan-European organizations primarily chose to focus on reform in more efficient ways closer to home, by lobbying directly to the relevant policy actors in Brussels. The second and more important reason is that there are few credible partners in the region with whom to work. The impact of Green 10 tutelage was most effective when it linked with established organizations in post-socialist countries. The only such example during our research was CEKOR from Novi Sad, which is an active member organization of CEE Bankwatch. However, more recently, a few organizations have joined Green 10 networks: CZZS (Banja Luka) joined Transport & Environment; Sredina – Association of Citizens (Belgrade) is a full member of EEB; CEPRO-MED (Belgrade) is a member of the Health and Environment

Alliance (HEAL); and Bird Study and Protection Society of Serbia (Novi Sad) is a member of BirdLife Europe.

What emerged quite clearly from our research was that the impact of Europeanization on the behaviour of actors and the success of social learning was contingent upon the interventions of IFIs and consultants who provided critical tutelage to state and ministry officials grappling with the new EIA process. Although we found evidence that such tutelage seems to have delivered a measure of progress, particularly the ability of relevant ministries to become competent authorities in steering EIA processes, the economic realities in Serbia and BiH dictate a high turnover of staff in the public sector and there is therefore a continual need to train individuals in the absence of strong institutional memory or stability.

Whilst the EU does offer technical assistance through Technical Assistance and Information Exchange (TAIEX), it targets harmonization, not social learning. In other words, the key difference between tutelage from the EU through TAIEX and the interaction between IFIs and local officials during EIAs is that the objective of the former is limited to teaching the beneficiary how to approximate EU legislation for formal compliance, whilst the latter is deployed to start the process to change long-standing practices and norms. As mentioned by one of the international technical consultants involved in the Corridor 5c EIA process, it was vital for those with technical expertise to understand that they were 'not just engineers anymore', and there was also an important social-environmental role to play. It is this second form of tutelage, if used effectively by the EU alongside other international actors, which can potentially trigger longer-term change in multi-level environmental governance in both BiH and Serbia.

Political futures?

The final set of conclusions that we need to draw from the research relates to what our findings may suggest about political transformation in the Western Balkans and in BiH and Serbia in particular. The advantage of undertaking empirical analysis of a particular policy/issue area is that it provides the opportunity to move beyond the discourse of state building and intervention, with its emphasis on external drivers and questions of sovereignty, towards an emphasis on the realities of domestic power and politics. This does not necessarily lead to more sanguine conclusions, but it does at least highlight instances of progressive change and allow for an explanation as to why positive change occurs and under which conditions. The initial question posed in the introduction to this book was *when Europeanization does deliver change, why does it happen, how does it happen, and which actors and institutions are critical?* In other words, our purpose has been to try to detect instances of progressive change and to better understand them.

From such a perspective a number of conclusions can be drawn from our research. In both countries, we have identified existent capacities and a commitment to political and governance reform. External assistance, whether at the behest of the EU or other donors is certainly not a panacea, but its impact is contingent on how it is delivered and which domestic actors it engages. What tutelage and assistance 'locks in' is critical – if such intervention benefits and empowers progressive coalitions or actors who happen to be located close to the source of political or bureaucratic power, then the impact is seemingly sustainable and transformative. If, on the other hand, there is a lack of contagion or there are insufficient partners with whom beneficiaries can work, then the impact is temporary or negligent.

How is this relevant for our understanding of the political constraints faced by these states? Serbia's progress towards EU membership has hitherto been determined by relations with Kosovo. Whilst this is unlikely to change entirely, the country's progress towards EU membership will now be also judged according to compliance with the *acquis* and the pace of progressive change. The extra-conditionality and the bespoke requirements are likely to be extremely demanding and costly, both economically and politically. Understanding how, why, and under what conditions Europeanization exerts a transformative power is therefore vital.

For BiH, the challenge is altogether different. What our findings suggest is that the constitutional framework and the political and bureaucratic inertia that has ensued since 1995 impede any degree of positive impact from Europeanization, beyond basic legal change. It was popular frustration with an intransigent and corrupt political class that prompted the protests in 2014; but it is also frustration with the EU and the lack of any progress towards accession that continues to fuel the discontent directed towards domestic elites and the international community. Getting the aid and assistance to work effectively is only part of the challenge; no amount of targeted capacity assistance, partnerships between donors and domestic actors, or bottom-up, on-the-ground aid will obviate the fundamental lack of regime change and the albatross of the Dayton constitution.

However, we do not see our findings as being limited to further confirmation of the continued inertia around constitutional reform and thus further EU integration in BiH, or the challenge of limited progress towards accession in Serbia due to the normalization of relations with Kosovo. The environmental optic provides ample coverage of this, but it also offers insight into a dimension of domestic grassroots politics that has yet to be properly analysed or fully understood. As a tantalizing post-script, the end of our research coincided with a proliferation of citizen mobilization throughout the Western Balkans, undoubtedly styled after the so-called Arab Spring and protests on city squares elsewhere in Europe, including major demonstrations in Greece, Spain, Bulgaria, and Slovenia. Particularly in Turkey and Croatia, the catalyst for citizen mobilization was planned construction of commercial properties on urban green space, and focused on the rights of residents to use public spaces, which also resonated with the (albeit smaller) 2005 Peti Park protests in Belgrade (Večernje novosti, 2008). Despite years of inertia, citizens of BiH followed suit. One particularly interesting case is that of 'Picin Park' in Banja Luka. In May 2012, the construction company, Grand Trade, began excavating the park in preparation for building commercial and residential properties. Citizens started protesting the development of the site, and over the course of the weeks that followed, local citizens from across the socio-economic spectrum gathered in the park along with local NGOs (Radio Slobodna Evropa, 2012). The protesters also organized a petition to mount a legal challenge to the actions by Grand Trade. As with Gezi Park in Istanbul, the environmental issues were vehicles of transition that exemplified wider social, political, and economic grievances in the country. Miodrag Dakić from CZZS was involved in the demonstrations, and he said that it was not just about a park but a 'pretext' to protest disaffection about corruption and exploitation by political and business leaders. He added that: 'Connecting people around a common goal is something that is very important, the construction of a common platform for co-operation, which is then easier to animate in the future. That is something that is very important in the long run' (Kelava, 2012).[2] Ultimately, the protests failed to save the park, but they may still have symbolic importance for future mass mobilization in RS. More significantly, in response to the closing of factories in Tuzla in FBiH, a number of mass protests started in a number of Bosnian cities in February 2014. Although there was a certain level of violence associated with this movement, much of the dissent soon consolidated around unofficial citizens' assemblies or plenums as a means of providing voice for their grievances. The plenums ceased to be active within a few months, leading one commentator to conclude that they had 'failed', but also conceding that: 'the protests and the plenums had a lasting impact. Another Bosnia is possible, and it is the responsibility of the people of Bosnia to build it. The plenums were an important first step' (Keil, 2014). In solidarity with protestors in BiH, mass demonstrations also occurred in Belgrade and Niš to show their support for social change in the neighbouring country (Frandino, 2014). The events over this period marked the first time since the end of the Bosnian wars in 1995 where citizens came out en masse to protest – not either in support of or against politicians, but rather against the characteristics of politics itself in the country. There is no established legacy of citizen protest during the transition period in BiH, since the post-conflict state came into being through an agreement amongst opposing combatants, which was brokered (and designed) internationally, and guaranteed by kin states that had exacerbated the violence. By contrast, although there has not been similar recent upheaval in Serbia, it has a stronger legacy of citizen mobilization, which dates back to the period leading up to the overthrow of the Milošević -led regime.

Linking the recent protests with the major road-building EIAs in BiH researched in this book, CEE Bankwatch issued the following warning to international actors (which also applies to Serbia):

> It would be foolish, though, to think the protest spirit has disappeared forever in BiH. And perhaps in this period of calm an institution such as the EBRD may pause to reflect on its role and past interventions in BiH society... it is important to mention the institution's lack of constructive communication with other stakeholders, including civil society organisations, regarding its activities in the country... If BiH is to survive and prosper, both its legislation and leaders must heed and respond to the needs of its population. The international community, including the EBRD, must help to ensure that this is realised, and support BiH citizens as they lead their own transition. (CEE Bankwatch, 2014)

Although directed at EBRD, it is also crucial for the EU to note the advice from CEE Bankwatch. In places like Serbia and BiH where fundamental socio-political transformation is necessary, the EU must move away from its predominant strategy of conditionality to achieve short-term guarantees on regional security, in favour of more bottom-up, inclusive intervention in other key policy areas, including the environment.

Like the construction of the pan-European Corridors that will eventually traverse the Western Balkans, building the road to Europe with respect to environmental governance will be gradual, arduous, and costly – but carefully creating the right foundations will benefit the citizens of BiH and Serbia, and provide permanent, mutually beneficial links with the rest of Europe.

Notes

Introduction

1. These two approaches are distinguished in Börzel (2005).

1 Europeanizing Environmental Governance

1. See Commission's 2001 paper on governance and also the Treaty of the European Union, Article 15.
2. In its White Paper on European Governance (2001), the European Commission posits that there are five necessary principles to achieve good governance at local, regional, national, or European levels: openness, accountability, effectiveness, participation, and coherence.
3. Interview with Communications Manager, T&E, 1 October 2009, Brussels; interview with Enlargement Co-ordinator, EEB, 2 October 2009, Brussels; interview with Programme Manager, WWF European Policy Office, 1 October 2009, Brussels.

2 The Curious Case of Bosnia-Herzegovina

1. In addition to CARDS, IPA also replaced Poland and Hungary Assistance for the Restructuring of the Economy (PHARE), Special Accession Programme for Agriculture and Rural Development (SAPARD), Instrument for Structural Policies for Pre-Accession (ISPA), and pre-accession assistance for Turkey.
2. The information can be found at the EU Delegation web page here: http://www.delbih.ec.europa.eu/Default.aspx?id=15&lang=EN (last accessed 7 August 2014).
3. IPA Multi-Annual Indicative Planning Document (MIPD) 2011–2013 for Bosnia and Herzegovina, p. 12, http://ec.europa.eu/enlargement/pdf/mipd_bih_2011_2013_En.pdf (last accessed 2 December 2013).
4. IPA MIPD 2011–2013 for Bosnia and Herzegovina, p. 19.
5. Interview with Project Officer, TACSO country office, Sarajevo, 4 June 2010.
6. Interview with Resident Adviser, TACSO country office. The list of LAG members can be found here: http://www.tacso.org/doc/TACSO%202%20BH%20LAG%20members%20for%20web-.pdf.
7. The agreement will be abbreviated as SAA in the notes for the remainder of this chapter.
8. SAA, Article 108.
9. After the provisional agreement by Dodik, Špirić returned to the Council of Ministers at the end of November 2007.

10. Interview with the Head of the Department for Environmental Protection, MOFTER Sarajevo, 13 May 2008.
11. Interview with the Head of the Department for Environmental Protection, MOFTER.
12. Interview with the Head of the Department for Environmental Protection, MOFTER.
13. Interview with the Head of the Department for Environmental Protection, MOFTER.
14. BiH has a three-person rotating presidency, with one representative from each constituent people.
15. Interview with the Head of the Department for Environmental Protection, MOFTER.
16. Interview with the Head of the Department for Environmental Protection, MOFTER.
17. Interview with an FMOIT official, Sarajevo, 20 October 2009.
18. Deputy Minister for Environmental Protection, Sarajevo Canton, Sarajevo, 19 February 2009.
19. Interview with an FMOIT official.
20. Interview with an official at the RS Ministry for Spatial Planning, Construction, and Environmental Protection, Banja Luka, 18 February 2009.
21. Interview with an official at the RS Ministry for Spatial Planning, Construction, and Environmental Protection.
22. Interview with the Head of the Department for Environmental Protection, MOFTER.
23. Interview with the Head of the Department for Environmental Protection, MOFTER.
24. Interview with Deputy Minister, Ministry for Spatial Planning and Environmental Protection, Sarajevo, 19 February 2009.
25. Interview with Deputy Minister, Ministry for Spatial Planning and Environmental Protection, Sarajevo.
26. In all of the Ministries except for Sarajevo Canton, the lead officials responsible for environmental permits were contacted to complete the questionnaire. In Sarajevo Canton, the corresponding official was on indefinite leave, so the questions were answered by the Deputy Minister.
27. Personal communication with official at the Ministry for Spatial Planning and Environmental Protection, Sarajevo Canton, October 2009.
28. Interview with official at the Ministry for Trade, Tourism, and Environmental Protection, Herzegovina-Neretva Canton, Mostar, 1 April 2009.
29. Interview with Deputy Minister, Ministry for Spatial Planning and Environmental Protection, Sarajevo.
30. Tuzla Canton did not respond to the question, but the other cantons indicated that they had not received any international donor assistance.
31. The municipalities are: Bužim (Canton 1); Domaljevac- Šamac (Canton 2); Čelić (Canton 3); Kakanj (Canton 4); Foča-Ustikolina (Canton 5); Donji Vakuf (Canton 6); Jablanica (Canton 7); Ljubuški (Canton 8); Ilijaš (Canton 9); and Drvar (Canton 10).
32. Personal communication with the Chair for the Board for urban planning, communal activities and environmental protection, RS, 11 May 2010.
33. Personal communication with the Chair for the Board for urban planning, communal activities and environmental protection, RS.

34. Interview with an Assistant to the Mayor, City of Sarajevo, 19 March 2010.
35. Interview with a Senior Expert Associate for Environmental Protection A, City of Mostar, 18 March 2010.
36. Interview with a Senior Expert Associate for Environmental Protection A, City of Mostar.
37. Interview with a Senior Expert Associate for Environmental Protection B, City of Mostar, 18 March 2010.
38. Personal communication with the Chair for the Board for urban planning, communal activities and environmental protection, RS.
39. Personal communication with an official at Trebinje Municipality, 27 May 2009.
40. In Čelić, officials were seeking to fill the vacancy for Municipal Inspector; the two posts for Municipal Inspectors in Bužim were vacant.
41. The City of Sarajevo was involved in the drafting of LEAP documents for its urban municipalities. The LEAP for Mostar had been drafted and was in the adoption phase.
42. Interview with the Director of REC (BiH), 4 November 2008, Sarajevo.
43. Interview with the Director of REC (BiH).
44. Interview with the Director of REC (BiH).
45. Interview with the President of Ekotim, 6 November 2008, Sarajevo.
46. Interview with the President of CZZS, 17 February 2009, Banja Luka.
47. The notable exception has been CZZS, which has had regional partnerships with the Serbian NGO Young Researchers of Serbia (MIS) and Green Action from Croatia for various projects since 2001. CZZS was also involved in the consultative Brussels-based Environmental Forum that discusses priorities for EU funding. There are Bosnian NGOs in regional CEE networks such as CEEWeb and ANPED.
48. Interview with the Communications Manager for T&E, 1 October 2009, Brussels.
49. Interview with the Co-ordinator for Enlargement, EEB, 2 October 2009, Brussels.
50. Interview with the Neighbourhood Programme Manager, European Policy Office, WWF, 1 October 2009, Brussels.

3 Environmental Impact Assessment (EIA) Processes in Bosnia-Herzegovina

1. The FBiH law is the Law on Environmental Protection, Official Gazette of FBiH, no. 33/03, 19 July 2003. The RS law is the Law on Environmental Protection (consolidated text), Official Gazette of RS, no. 28/07, 28 March 2007.
2. Pravilnik o pogonima i postrojenjima za koje je obavezna procjena utjecaja na okoliš i pogonima i postrojenjima koji mogu biti izgrađeni i pušteni u rad samo ako imaju okolinsku dozvolu (Official Gazette of FBiH, no. 19/04, 10 April 2004), Art. 3.
3. Official Gazette of FBiH, no. 04/19, Art. 2 and Art. 6.
4. The responsible ministry for larger projects is the Ministry for Environment and Tourism (FMOIT) in FBiH, and the Ministry for Urban Planning, Construction and Ecology in RS.

166 Notes

5. Personal communication with an official in the Environmental Department, RS Roads Directorate.
6. Unless indicated, the information from this sub-section is drawn from the English-language translation of the Corridor Vc Motorway Project: Environmental Impact Assessment Study, available here from the EBRD website: http://www.ebrd.com/english/pages/project/eia/38716.pdf. Last accessed 7 August 2014.
7. Interview with an official in the Department of Environmental Protection, Ministry for Physical Planning, Civil Engineering and Ecology, RS, Banja Luka, 18 February 2009.
8. Corridor Vc Motorway Project: Environmental Impact Assessment Study.
9. The document was prepared through consultation and co-operation between the World Bank, IMF, state-level Council of Ministers, the entity-level governments, and the Office of the BiH Co-ordinator for Preparation of PRSP.
10. The description of the project is from the July 2009 EIA study provided to stakeholders before the public hearing on the road project.
11. Personal communication with a Stolac municipal official.
12. MAC BiH is a three-person technical body (one from each constituent people) under the Ministry of Civil Affairs of BiH and is responsible for the identification and marking of the threat level of places due to landmines.
13. The Commission was created in Annex 8 of the Dayton-Paris Agreement (1995) as a means of preserving cultural heritage in BiH. The Commission (composed of local representatives from each constituent people and international members) designates sites of having special cultural or historic significance.
14. For more details, see: http://whc.unesco.org/en/tentativelists/5280/
15. For more details, see: http://whc.unesco.org/en/tentativelists/5092/
16. Corridor Vc Motorway Project: Environmental Impact Assessment Study, p 13–14.
17. Corridor Vc Motorway Project: Environmental Impact Assessment Study, p. 21.
18. Interview with official (1) at the Department of Environmental Protection, Ministry for Physical Planning, Civil Engineering and Ecology, RS, Banja Luka, 18 February 2009.
19. Personal communication from a former official at the RS Roads Directorate, 16 March 2009.
20. Banja Luka-Gradiška Road Scoping Report, February 2004. Provided by the RS Roads Directorate in February 2009.
21. Banja Luka-Gradiška Road Scoping Report, Appendix 3.
22. Interview with official in the Environment Department, RS Roads Directorate, Banja Luka, 17 February 2004.
23. Interview with official (2) at the Department of Environmental Protection, Ministry for Physical Planning, Civil Engineering and Ecology.
24. Personal communication from a former official at the RS Roads Directorate.
25. Interview with official at EBRD, Sarajevo, 19 February 2009.
26. Personal communication from a former official at the RS Roads Directorate.
27. Interview with the Centre for Environment (CZZS), Banja Luka, 12 May 2008.
28. Interview with the President of CESD, Sarajevo, 4 November 2008.

29. Interview with REC (Banja Luka office), Banja Luka, 5 November 2008.
30. Telephone interview with official at ERM (Germany), 23 March 2009.
31. Interview with an official at the Federal Roads Directorate, Sarajevo, 19 February 2009.
32. Telephone interview with official at ERM (Germany).
33. Interview with an official at the Federal Roads Directorate, Sarajevo.
34. Public hearing on the EIA for the Mostar Bypass, Rodoč, 2 September 2009.
35. Public hearing on the EIA for the Mostar Bypass, Rodoč.
36. Public hearing on the EIA for the Buna-Neum Road, Neum, 12 January 2010.
37. Public hearing on the EIA for the Buna-Neum Road, Neum; public hearing on the EIA for the Buna-Neum Road, Stolac, 13 January 2010.
38. Public hearing on the EIA for the Mostar Bypass, Rodoč.
39. Public hearing on the EIA for the Buna-Neum Road, Neum; public hearing on the EIA for the Buna-Neum Road, Stolac.
40. Interview with the Deputy Minister for Environmental Protection, Ministry for Spatial Planning and Environmental Protection, Sarajevo Canton, Sarajevo, 19 February 2009.
41. Interview with the Minister, Ministry of Construction and Urban Planning, Herzegovina-Neretva Canton (HNK), Mostar, 2 June 2010.
42. Interview with the Minister, Ministry of Construction and Urban Planning, HNK.
43. Interview with the Senior Experts Associates, Department of Public Utilities and Environmental Protection, City of Mostar, 18 March 2010.
44. Public hearing on the EIA for the Mostar Bypass, Rodoč, 2 September 2009. They identified themselves as representatives from Construction Regulations, Department for Infrastructure, and Department for Urban Planning at the City of Mostar.
45. Public hearing on the EIA for the Buna-Neum Road, Neum.
46. Public hearing on the EIA for the Buna-Neum Road, Stolac.
47. Public hearing on the EIA for the Buna-Neum Road, Stolac.
48. The Ministers for Justice, Finance, Environment, and Agriculture all resigned at this time.
49. Interview with EBRD official, London, 15 October 2009.
50. More information about the WBIF can be found here: http://www.wbif.eu/About/. Last accessed 7 August 2014.
51. The 2003 EBRD environmental policy document is available online here: http://www.ebrd.com/downloads/research/policies/policy.pdf. Last accessed 7 August 2014.
52. Interview with EBRD official, Sarajevo, 19 February 2009.
53. Interview with an official at the FBiH Roads Directorate, 9 October 2009.
54. Interview with an official at the FBiH Roads Directorate.
55. Interview with EBRD official, Sarajevo.
56. Interview with EBRD official, written response from a former official at the RS Roads Directorate, 16 March 2009.
57. Interview with an official at the FBiH Roads Directorate, 9 October 2009.
58. Public hearing on the EIA for the Mostar Bypass, Rodoč.
59. Written correspondence from officials at BCEOM, 8 May 2009.

168 Notes

60. Written comments from an official at Louis Berger, 15 April 2009.
61. For a description of Louis Berger's experience in the Western Balkans, see: http://www.bergergroup.com/?p=open_work&type=work&id=42&t=The_Balkans
62. For more details on the time line for the Sava Bridge, see: http://www.savabridge.com/progress.htm
63. For a description of the Sava Bridge construction project, see: http://www.savabridge.com/description.htm
64. Written comments from an official at Louis Berger.
65. Written comments from an official at ERM, 23 March 2009.
66. Written comments from an official at ERM.
67. Interview with an official at the FBiH Roads Directorate.
68. Interview with an official at the FBiH Roads Directorate.
69. PCDP for the Banja Luka-Gradiška EIA, included as an annex to the scoping report.
70. Interview with an official from CZZS, Banja Luka, 17 February 2009.
71. Interview with an official from CZZS, Banja Luka.
72. Interview with a project manager at the REC project office, Banja Luka, 12 May 2008.
73. Interview with an official from CZZS, Banja Luka, 17 February 2009.
74. Telephone interview with an official from Ekotim, 19 March 2009. The interviewee suggested university professors establish NGOs as grant-seeking organizations, since they may be eligible for money as non-profit bodies which they would not be as academic bodies.
75. Personal communication with an officer from Fondeko, 4 March 2009.
76. Interview with an official at CESD, Sarajevo, 4 November 2008.
77. Interview with an official at CESD, Sarajevo, 13 February 2008.
78. Written correspondence from an official at the Soil Science Institute, Sarajevo, 6 March 2009.
79. Interview with an official from Ekotim, Sarajevo, 6 November 2008.
80. Public hearings for the Buna-Neum road EIA in Neum and Stolac, 12–13 January 2010.
81. Citizens during a focus group organized by the researchers after the Mostar Bypass EIA public meeting in Rodoč, 2 September 2009. There were four citizens who were quite active at the public hearing; they are indicated as C1–C4 in the following notes.
82. Comments by an FMOIT official after the Buna-Neum EIA meeting in Stolac, 13 January 2010.
83. Comments made by a Novi Val representative at a focus group in Blagaj, 18 March 2010.
84. Interview with an official from Ekotim, Sarajevo, 9 October 2009; interview with an official from CZZS, Banja Luka, 17 February 2009.
85. Written comments by a former official at the RS Roads Directorate, 16 March 2009; interview with an official at EBRD, Sarajevo.
86. Telephone interview with an official at ERM.
87. Comments by citizen (C1) during the focus group, Rodoč, 2 September 2009.
88. Comments by citizen (C4) during the focus group, Rodoč, 2 September 2009.

89. Comment by a citizen during the Buna-Neum EIA public meeting, Stolac.
90. Literally translated 'local community'; it is a locally operated neighbourhood organization.
91. The researchers distributed a questionnaire at the Mostar Bypass public hearing. There were approximately 30 attendees, of whom 14 completed the questionnaire. The head of the *mjesna zajednica* assisted the research team in distributing the surveys.
92. Written communication from an official at CEE Bankwatch, 10 March 2011.
93. Interview with an official from Ekotim, Sarajevo, 9 October 2009.
94. Focus group with coalition members in Blagaj, 18 March 2010.
95. Written communication from an official from CZZS, 27 September 2009.
96. Comments by the head of the coalition in Blagaj during a focus group, 18 March 2010.
97. Comment by the former mayor of Mostar South-east (Blagaj), during a focus group, 18 March 2010.
98. Interview with a member of the Association of Local Citizens of Počitelj, Mostar, 3 June 2010.
99. Comment by the owner of the Jaffa-Komerc fruit orchards, Blagaj, during a focus group, 18 March 2010.
100. Interview with a member of the Association of Local Citizens of Počitelj.

4 From Pariah to Partner? The Case of Serbia

1. This evolution from 'pariah' to 'partner' is coined by De Launey (2014).
2. Slobodan Milošević founded the Socialist Party of Serbia (SPS) in 1990, and was President of Serbia during the wars in Croatia and BiH, and President of the Federal Republic of Yugoslavia during the war in Kosovo.
3. A full description of the 2013 country-based EIDHR assistance in Serbia can be found here: http://www.europa.rs/sw4i/download/files/article/guidelines%20for%20aplicants.pdf?id=951. Last accessed 23 September 2014.
4. More information can be found here: http://www.europa.rs/en/eu_assistance_to_serbia/ipa/IPA_2013.html. Last accessed 8 August 2014.
5. The list of individuals on the LAG can be found here: http://www.tacso.org/doc/TACSO%202%20Serbia%20LAG%20members%20for%20web.pdf. Last accessed 23 September 2014.
6. A timeline of the history of the CTF and the Enhanced Permanent Dialogue, and Serbia–EU relations more generally, can be found on the website of the Serbian European Integration Office (SEIO): http://www.seio.gov.rs/serbia-and-eu/history.60.html. Last accessed 8 August 2014.
7. The SEIO listed these areas: 'internal market, competition, transport, energy, nuclear safety, information society, network of electronic communications, audio visual sector, justice, liberty and security, trade policy, industrial policy and SME, customs and taxation, fiscal and monetary policy, property rights, as well as environmental protection, food safety and financial control'.

8. The timetable of the 2010 EPD meetings can be found here: http://www.seio.gov.rs/upload/documents/Izvestaji/planning_%20serbia_%202010.pdf
9. SAA, Article 135, para. 2.
10. The SAA was unfrozen in June 2010, whilst the interim trade agreement had been unfrozen in December 2009.
11. For Serbia's replies to the questionnaire, see: http://www.seio.gov.rs/upload/documents/upitnik/srb_questionnnaire_Engl.pdf
12. A detailed list of competencies can be found here: http://www.seio.gov.rs/office/about-us.58.html. Last accessed 8 August 2014.
13. Detailed information about competencies for each department within SEIO can be found here: http://www.seio.gov.rs/office/departments.59.html. Last accessed 8 August 2014.
14. Interview with the Head of Group for Environment, Agriculture and Rural Development, SEIO, Belgrade, 3 November 2009.
15. Interview with the Head of Group for Environment, Agriculture and Rural Development, SEIO, Belgrade.
16. Interview with the Head of Group for Environment, Agriculture and Rural Development, SEIO, Belgrade.
17. Interview with the Biodiversity Project Officer, IUCN, Belgrade, 1 December 2009.
18. Interview with the Deputy Director, BCIF, Belgrade, 4 February 2010.
19. Interview with representative of Environmental Ambassadors, Belgrade, 4 November 2009. Interview with the General Secretary of the Young Researchers of Serbia, Belgrade, 5 November 2009.
20. Interview with representative of Environmental Ambassadors, Belgrade. However, the positive assessments of Dulić were made before his indictment and sentencing for abuse of power related to issuing permits to the Slovenian company Nuba Invest (Beta & Tanjug, 2013).
21. Interview with the Project Manager, BCIF, Belgrade, 4 February 2010.
22. The National Assembly adopted the following laws: Law on Waste Management; Law on Packaging and Repackaging Waste; Law on Noise Protection; Law on Biocides; Law on Air Protection; Law on Nature Protection; Law on Sustainable Fishery Use; Amendments to the Law on the Environmental Impact Assessment; Law on Chemicals Amendments and Supplements to the Law on Environmental Protection; Law on ratification of Belarusian amendment to Annex B of the Kyoto Protocol to the United Nations Framework Convention on Climate Change; Law on ratification of Convention Access to Information, Public Participation in Decision-making and legal protection in matters of environmental protection (Aarhus Convention); Law on ratification of on the procedure of giving consent to the notice of certain hazardous chemicals and pesticides in international trade (Rotterdam Convention); Law on Protection of non-ionizing radiation; Law on protection against ionizing radiation and nuclear safety; and the Law on the Prohibition of the Development, Production, Stockpiling and Use of Chemical Weapons and on Their Destruction.
23. Further information about the approval procedure for projects in Serbia under CDM of the Kyoto Protocol can be found on the website of the Designated National Authority: http://www.ekoplan.gov.rs/DNA/index_En.html.

24. Interview with an Adviser at the Ministry for Environmental Protection and Spatial Planning, Belgrade, 4 November 2009.
25. Interview with the Head of Group for Environment, Agriculture and Rural Development, SEIO, Belgrade.
26. For more details, see: http://www.sepa.gov.rs/index.php?menu=2&id=-4&akcija=showAll. EEA is an agency operating at the European level to collect data and provide assessments for the benefit of its members (the 27 EU countries, Liechtenstein, Switzerland, Turkey, Iceland and Norway), as well as six co-operating countries from the Western Balkans (BiH, Serbia, Croatia, Macedonia, Montenegro, and Albania). EIONET is a network of the member and co-operating countries, which co-ordinated by the EEA.
27. Law on Nature Protection (2004), Art. 92.
28. Interviews with the Assistant Director and an Adviser at the Institute for Nature Protection, Belgrade, 1 December 2009.
29. See: http://www.zzps.rs/novo/index.php?jezik=en&strana=o_nama_delatnost
30. Interview with an Adviser at the Institute for Nature Protection, Belgrade.
31. Statute of the Autonomous Province of Vojvodina, Article 26. Ruthenian is an Eastern Slavic language, referred to as *rusinski* in Serbian.
32. Omnibus Law, Article 29.
33. Interview with the Assistant to the Director, Provincial Secretariat for Environmental Protection, Vojvodina, Novi Sad, 2 December 2009.
34. Interview with the Assistant to the Director, Provincial Secretariat for Environmental Protection, Vojvodina. The Assistant to the Director also said that the Fruška Gora National Park had been the responsibility of Vojvodina until 1990, and the Province would thus be 'getting back' competence.
35. Interview with the Assistant to the Director, Provincial Secretariat for Environmental Protection, Vojvodina.
36. Interview with the Secretary of the Committee on Environmental Protection, Standing Conference of Towns and Municipalities, Belgrade, 3 February 2010.
37. Interview with an official at the Secretariat for Environmental Protection, City of Belgrade, 1 December 2009.
38. Personal communication from an official from the urban municipality of Crveni Krst, 28 April 2010.
39. Interview with the Head of Department for EIA, Ministry for Environment & Spatial Planning, Serbia, Belgrade, 4 November 2009.
40. Interview with the Head of Group for Environment, Agriculture and Rural Development, SEIO.
41. Interview with an official at Environmental Ambassadors, Belgrade, 4 November 2009.
42. Interview with the Vice-President of Ecotopia, Belgrade, 5 February 2010.
43. Interview with the General Manager of Eco-libri, Belgrade, 6 November 2009.
44. Interview with an official at Environmental Ambassadors, Belgrade.
45. Interview with an official from Corridor 10 LLC, Belgrade, 30 November 2009.
46. Interview with an official at Eko Centar, Belgrade, 3 November 2009.
47. Interview with an official at the Green Network of Vojvodina, Novi Sad, 2 December 2009.

172 Notes

48. Interview with the General Secretary of Young Researchers of Serbia, Belgrade, 5 November 2009.
49. Interview with officials at SKGO, Belgrade, 3 February 2010.
50. The historical background of the SKGO can be found here: http://www.skgo.org/pages/display/4/History
51. Interview with officials at SKGO.
52. Interview with officials at SKGO.
53. In addition to Batočina, the 33 other members of the Committee for Environmental Protection for the session starting in 2010 are: Ada, Bajina Bašta, Barajevo, Bačka Palanka, Bela Crkva, Bečej, Beočin, Bor, Bojnik, Bujanovac, Veliko Gradište, Vlasotince, Vladimirci, Vrnjačka Banja, Dimitrovgrad, Zemun (Beograd), Ivanjica, Kosjerić, Koceljeva, Kruševac, Kraljevo, Lazarevac (Beograd), Lebane, Medijana (Niš), Niš, Obrenovac (Beograd), Pančevo, Rača, Sečanj, Surčin (Beograd), Smederevo, Surdulica, and Sabač.
54. Interview with officials at SKGO. Batočina is a small municipality in Central Serbia. The Mayor of Batočina represents the municipality as Chairman.
55. Translated transcript of SKGO Committee for Environmental Protection, Veliko Gradište, 24 February 2010.
56. Translated transcript of SKGO Committee for Environmental Protection.
57. Translated transcript of SKGO Committee for Environmental Protection.
58. Translated transcript of SKGO Committee for Environmental Protection.
59. Translated transcript of SKGO Committee for Environmental Protection. The representative from Koceljeva also joined the working group.
60. Translated transcript of SKGO Committee for Environmental Protection.
61. This includes the non-response letter from Crveni Krst, an urban municipality in Niš. The 44 local administrations that responded were: Bačka Topola, Bajina Bašta, GO Barajevo, Bela Crkva, Bela Palanka, Beočin, Bojnik, Bor, Bujanovac, Dimitrovgrad, Ivanjica, Koceljeva, Kosjerić, Kragujevac, Kraljevo, Kruševac, GO Lazarevac, Majdanpek, Mali Idjoš, GO Medijana, Medvedja, Mionica, Niš, Nova Crnja, Novi Sad, GO Obrenovac, Opovo, Pančevo, Paraćin, Prokuplje, Rača, Sabač, Sečanj, Smederevo, Stara Pazova, GO Surčin, Surdulica, Topola, Trstenik, Vladičin han, Vlasotince, Vrnjačka Banja, Žabalj, and GO Zemun. The gradske opštine (GO) or urban municipalities are indicated by GO. The cities that submitted replies were: Kragujevac, Kraljevo, Kruševac, Niš, Novi Sad, Pančevo, Sabač, and Smederevo.
62. Interview with the Head of City Administration for Environmental Protection, Novi Sad, 2 December 2009.
63. Interview with the Head of City Administration for Environmental Protection, Novi Sad.
64. Interview with the Head of City Administration for Environmental Protection, Novi Sad; questionnaire response.
65. From questionnaire response.
66. Interview with the Head of City Administration for Environmental Protection, Novi Sad.
67. Interview with an EIA Specialist, City Administration for Environmental Protection, Novi Sad, 2 December 2009.
68. Interview with the Head of City Administration for Environmental Protection, Novi Sad.
69. Interview with an official at the City of Belgrade, 1 December 2009.

70. Kragujevac, Kraljevo, Kruševac, Niš, Novi Sad, Pančevo, Sabač and Smederevo.
71. Questionnaire response. The urban municipality also had a separate department, with one staff member and a budget of 300,000 RSD in 2009.
72. Questionnaire response.
73. Questionnaire response.
74. Interview with officials at SKGO.
75. The list of criteria was provided by a Project Management Assistant, USAID, Belgrade office.
76. Interview with a Project Management Assistant, Economic Growth Office, USAID, Belgrade, 3 December 2009.
77. Interview with an Investment Project Specialist, The Urban Institute, Belgrade, 3 December 2009. The consultant from the Urban Institute co-ordinated the MEGA programme and was also involved in MISP.
78. Interview with an Investment Project Specialist, The Urban Institute.
79. Interview with an Investment Project Specialist, The Urban Institute.
80. Interview with an Investment Project Specialist, The Urban Institute.
81. Interview with a Project Management Assistant, Economic Growth Office, USAID, Belgrade.
82. Interview with representatives of the European Movement in Serbia, Belgrade, 4 February 2010.
83. Interview with representatives of the European Movement in Serbia.
84. Interview with a Programme Manager, REC Serbia Country Office, Belgrade, 3 November 2009.
85. Interview with the State Secretary, Ministry for Human and Minority Rights, Republic of Serbia, 1 December 2009.
86. Interview with Project Management Specialist, Civil Society, USAID, Belgrade, 3 December 2009.
87. Interview with the Head of Mission, ISC Serbia.
88. Interview with the Vice-President of Ecotopia.
89. Interview with the General Secretary, Young Researchers of Serbia.
90. Interview with a Programme Manager, REC Serbia Country Office.
91. Interview with the State Secretary, Ministry for Human and Minority Rights.
92. Interview with the General Secretary, Young Researchers of Serbia.
93. Interview with the Interim Director, IUCN, Belgrade.
94. Interview with a Programme Manager, REC Serbia Country Office.
95. Interview with a Programme Manager, REC Serbia Country Office.
96. Interview with an Investment Project Specialist, The Urban Institute, Belgrade.
97. Interview with the Head of Mission, ISC Serbia.
98. Interview with the Vice-President of Ecotopia.
99. Interview with the Interim Director of IUCN, Belgrade.
100. Interview with the Green Network of Vojvodina, 2 December 2009.
101. Interview with the Director of the Ecological Movement of Novi Sad.
102. Interview with the President of Serbian Green Youth, 4 November 2009.
103. Interview with the President of Ekotim.
104. Interview with the Young Researchers of Serbia.
105. Interview with the Young Researchers of Serbia.

174 Notes

106. The 12 organizations are: Bela breza, Kruševac; Ekološki klub 'Zeleni putokazi', Raška; Biciklističko pešačko udruženje 'Lisac', Žagubica; Zeleni ključ, Niš; Odred izviđača 'Josif Pančić', Niš; Green Bike, Valjevo; Porečje, Vučje; Odred izviđača 'Zavičaj 1903', Vranje; Moravski orašak, Trstenik; Centar za održivu budućnost 'Zelena zona', Knjaževac; Odred izviđača 'Šumatovac', Aleksinac; and GM Optimist, Gornji Milanovac. With the help of BCIF, a number of the 25 organizations completed an electronic questionnaire. The information from these surveys will be used here.
107. Many of these organizations also indicated that education and training were an important component of their activities.

5 Environmental Impact Assessment (EIA) Processes in Serbia

1. The EIA legislation has been amended, most recently in 2009. The law amending the law on the assessment of environmental impact (36/09) was ratified in September 2009.
2. The government has a list of projects for which EIA is required; the Ministry determines whether an EIA study is needed for other types of projects.
3. This information should be accompanied information on: spatial/urban planning, preliminary design, maps, and proof of approval by relevant bodies.
4. The required information that the developer needs to include is outlined in Article 17 of the legislation.
5. To streamline the EIA process, Article 21 of the legislation has been deleted in the 2009 amendments. In the deleted part of the law, the competent authority would collect the feedback from the consultation and send a report to the developer within 15 days of the public hearing. The developer would then have a further 15 days to resubmit the EIA study, addressing the comments sent by the competent authority. Once this revised EIA is received, the competent authority has 15 days to send the EIA study and report on the public consultation to the Technical Commission. Because of this streamlining, an additional Article 22a has been added that stipulates that the competent authority must submit the public consultation report and EIA study to the Technical Commission within three days of the public meeting.
6. Corridor 10 Highway Project, Preliminary Design for E-80 Highway, NIŠ – DIMITROVGRAD, Section Prosek-Bulgarian border, REVISED CORRIDOR LEVEL EIA REPORT (November 2010). Page numbers in this sub-section refer to this document.
7. The lengths of the sub-sections of the road can be found here: http://www.koridor10.rs/index.php?mod=static&id=1106
8. Corridor 10 Highway Project Preliminary Design for E-75 Highway, NIŠ – border of FYRM, Section: Grabovnica-border of FYRM (hereafter PEIA E-75). Page numbers in this sub-section refer to this document.
9. A further two sections of the road were not included in the preliminary design and are not included in the study: Levosoje – Bukurevac; and Bukurevac – border of Macedonia. The lengths of the five sections included are available from the website of the developer, Corridors of Serbia: http://www.koridor10.rs/index.php?mod=static&id=1106

Notes 175

10. The full EIA is not available online, but an executive summary of the March 2005 feasibility study can be found on the EBRD website and was used for the project summary in this sub-section. URL: http://www.ebrd.com/pages/project/eia/36651e.pdf
11. The Master Plan for Belgrade can be found here: http://www.beograd.rs/cms/view.php?id=201126
12. Interview with a consultant from the Corridor 10 (K10) LLC, Belgrade, 4 November 2009; interview with a World Bank official, Belgrade, 23 February 2010.
13. Interview with a World Bank official, Belgrade.
14. Interview with a World Bank official, Belgrade.
15. Interview with a World Bank official, Belgrade.
16. Interview with the Expert Associate for Environmental Protection, PE Roads of Serbia.
17. Interview with the Expert Associate for Environmental Protection, PE Roads of Serbia.
18. Interview with a World Bank official, Belgrade.
19. Interview with an EBRD official, London, 15 October 2009.
20. Interview with the Expert Associate for Environmental Protection, PE Roads of Serbia.
21. Interview with the Head of the Environmental Section, Highway Institute, Belgrade, 30 November 2009.
22. Interview with an official responsible for EIA, Environmental Secretariat of Vojvodina, Novi Sad, 2 December 2009.
23. Interview with an official responsible for EIA in Vojvodina. The official also surmised that the reason why there may have been low levels of local interest is that there is a road along the existing alignment for E-75, so there will not be as significant a change when the road is upgraded compared to parts of Corridor 10 where there will be new road construction.
24. Interview with the Assistant to the Director, Environmental Secretariat of Vojvodina, Novi Sad, 2 December 2009.
25. Interview with an Adviser, Institute for Nature Protection of Serbia, Belgrade, 1 December 2009.
26. Interview with the Assistant Director, Institute for Nature Protection of Serbia, Belgrade, 1 December 2009.
27. Interview with an Adviser, Institute for Nature Protection of Serbia.
28. Interview with the Assistant Director, Institute for Nature Protection of Serbia.
29. Interview with an Adviser, Institute for Nature Protection of Serbia.
30. The Ministry has since been renamed the Ministry for Environment, Mining, and Spatial Planning.
31. Interview with the Head of Department for EIA, Ministry for Environment and Spatial Planning, Republic of Serbia, Belgrade, 4 November 2009.
32. The documents can be found here on the World Bank website: http://web.worldbank.org/external/projects/main?pagePK=51351038&piPK=51351152&theSitePK=40941&projid=P108005
33. The documents can be found here on the Corridors of Serbia (formerly Corridor 10) website: http://koridor10.rs/zashtita-zhivotne-sredine

176 Notes

34. Interview with a consultant from the Corridor 10 (K10) LLC, Belgrade, 4 November 2009.
35. Interview with the Head of the Environmental Section, Highway Institute.
36. Interview with a consultant from the Corridor 10 (K10) LLC, 4 November 2009.
37. Interview with the Expert Associate for Environmental Protection, PE Roads of Serbia, Belgrade, 25 February 2010.
38. Interview with a consultant from the Corridor 10 (K10) LLC, 30 November 2009.
39. Interview with a consultant from the Corridor 10 (K10) LLC, 30 November 2009.
40. Interview with a consultant from the Corridor 10 (K10) LLC, 30 November 2009.
41. Interview with an EBRD official, London.
42. Interview with a consultant from the Corridor 10 (K10) LLC, 4 November 2009.
43. Interview with the Expert Associate for Environmental Protection, PE Roads of Serbia.
44. The following paragraph is a summary of the views during the interview with a consultant from the Corridor 10 (K10) LLC, 4 November 2009.
45. The EMP documentation is outlined on the World Bank website here: http://web.worldbank.org/WBSITE/EXTERNAL/PROJECTS/EXTPOLICIES/EXTOPMANUAL/0,,contentMDK:20065218~menuPK:64701637~pagePK:64709096~piPK:64709108~theSitePK:502184,00.html
46. Interview with a consultant from the Corridor 10 (K10) LLC, 30 November 2009.
47. Interview with an engineer at the Centre for Roads of Vojvodina, Novi Sad, 2 December 2009.
48. Interview with an engineer at the Centre for Roads of Vojvodina.
49. The history of HIP can be found here: http://www.highway.rs/eng/history.html
50. The history of CIP can be found here: http://www.sicip.co.rs/ci/oNama/istorijat.html
51. Interview with an EBRD official, London.
52. Interview with an EBRD official, London.
53. Interview with a member of the environmental department at the Highway Institute, Belgrade, 30 November 2009. On this part of Corridor 10, CIP is the lead for the environmental assessment between Pirot and Dimitrovgrad.
54. Interview with a member of the environmental department at the Highway Institute, Belgrade.
55. Interview with a consultant from the Corridor 10 (K10) LLC, 30 November 2009. For the EMP, if there is no detailed design, the technical consultant uses the preliminary design.
56. Interview with a member of the environmental department at the Highway Institute, Belgrade.
57. Interview with a World Bank official, Belgrade.
58. Interview with an EBRD official, London.
59. The EIB response (17 June 2009) can be found on the CEE Bankwatch website: http://bankwatch.org/documents/response_EIB_Gazela_17Jun09.pdf

60. Interview with a World Bank official, Belgrade.
61. Interview with an EBRD official, London.
62. Interview with a World Bank official, Belgrade.
63. Interview with an EBRD official, London.
64. Interview with an official at the Department for EIA, Ministry for Environment and Spatial Planning, Belgrade.
65. Interview with the Head of the Environmental Section, Highway Institute.
66. Interview with an official at the Department for EIA, Ministry for Environment and Spatial Planning, Belgrade.
67. Interview with the Head of the Environmental Section, Highway Institute.
68. Interview with an official responsible for EIA, Provincial Secretariat for Environmental Protection, Vojvodina, Novi Sad, 2 December 2009.
69. Personal communication with an official at the World Bank office in Belgrade, 7 October 2013.
70. Corridor-level EIA report, vol. 2 (May 2009): Niš-Dimitrovgrad E-80, section Prosek-Bulgarian border.
71. EMP and Checklist E-75 Niš-FYROM border, section Grabovnica-Grdelica (July 2011, final: 21 September 2011).
72. EMP and Checklist E-75, Niš-FYROM border, section Manajle Tunnel-Vladičin Han (April 2014, final: 14 April 2014).
73. Corridor-level EIA report, vol. 2 (May 2009): Niš-Dimitrovgrad E-80, section Prosek-Bulgarian border.
74. Corridor level EIA, E-80, section Prosek-border of Bulgaria (final: November 2010).
75. Corridor-level EIA report, vol. 2 (May 2009): Niš-Dimitrovgrad E-80, section Prosek-Bulgarian border.
76. EMP and Checklist E-75 Niš-FYROM border, section Grabovnica-Grdelica (July 2011, final: 21 September 2011); Corridor-level EIA report, vol. 2 (May 2009): Niš-Dimitrovgrad E-80, section Prosek-Bulgarian border.
77. EMP and Checklist E-75, Niš-FYROM border, section Manajle Tunnel-Vladičin Han (April 2014, final: 14 April 2014)
78. EMP and Checklist E-75 Niš-FYROM border, section Grabovnica-Grdelica (July 2011, final: 21 September 2011).
79. EMP and Checklist E-75, Niš-FYROM border, section Manajle Tunnel-Vladičin Han (April 2014, final: 14 April 2014).

Conclusion

1. Of the Western Balkan territories, Albania and Croatia became full members of NATO in 2009. Macedonia has also applied, but its membership has been blocked by Greece over a naming dispute – Greece claims that keeping the name 'Macedonia' has irredentist connotations related to the Greek region of the same name.
2. This article also underlined how the Picin Park campaign was inspired by earlier protests over rights to public spaces in Croatia.

Bibliography

Amnesty International. (2010a). Serbia Must End Forced Evictions of Roma [Press release]. Retrieved from http://www.amnesty.org/en/news-and-updates/report/serbia-must-end-forced-evictions-roma-2010-06-10

Amnesty International. (2010b). Serbia: Vicious Circle of Forced Evictions in Serbia [Press release]. Retrieved from http://www.amnesty.org/en/library/asset/EUR70/012/2010/en/6f54ed41-dd86-43f8-94a6-8b171acdcbee/eur700122010en.html

Andonova, L. B. (2003). *Transnational Politics of the Environment: The European Union and Environmental Policy in Central and Eastern Europe.* Cambridge, Mass.: London: MIT.

Andric, G. (2010, 4 December). Belgrade's silent killer, *Balkan Insight.* Retrieved from http://www.balkaninsight.com/en/article/belgrade-s-silent-killer

AP. (2007, 2 December). Bosnian police reform deal in sight?, *B92.* Retrieved from http://www.b92.net/eng/news/region.php?yyyy=2007&mm=12&dd=02&nav_id=45869

B92. (2007, 24 October). Bosnian political crisis deepens, *B92.* Retrieved from http://www.b92.net/eng/news/region.php?yyyy=2007&mm=10&dd=24&nav_id=44825

B92. (2008, 15 October). Vojvodina assembly adopts statute, *B92.* Retrieved from http://www.b92.net/eng/news/politics.php?yyyy=2008&mm=10&dd=15&nav_id=54237

B92. (2014, 12 November). Šešelj returns to Serbia, addresses supporters, *B92.* Retrieved from http://www.b92.net/eng/news/politics.php?yyyy=2014&mm=11&dd=12&nav_id=92219

B92, Beta, & Tanjug. (2008, 29 April). Serbia, EU sign SAA in Luxembourg, *B92.* Retrieved from http://www.b92.net/eng/news/politics.php?yyyy=2008&mm=04&dd=29&nav_id=49819

B92, Beta, & Tanjug. (2009, 24 November). Parliament begins debate on Vojvodina statute, *B92.* Retrieved from http://www.b92.net/eng/news/politics.php?yyyy=2009&mm=11&dd=24&nav_id=63256

B92, Beta, Tanjug, VoA, & Deutsche Welle. (2008, 22 July). EU: Karadžić arrest important step, *B92.* Retrieved from http://www.b92.net/eng/news/politics.php?yyyy=2008&mm=07&dd=22&nav_id=52102

Bäckstrand, K. (2006). Multi-stakeholder partnerships for sustainable development: rethinking legitimacy, accountability and effectiveness. *European Environment, 16*(5), 290–306.

Baker, G. B. (1998). Civil society and democratisation theory: an inter-regional comparison. [electronic resource].

Baker, G. B. (1999). The taming of the idea of civil society. *Democratization, 6*(3), 1–29.

BBC. (2000a, 23 December). Serb reformists scent landslide victory, *BBC News.* Retrieved from http://news.bbc.co.uk/1/hi/world/europe/1083694.stm

Bibliography 179

BBC. (2000b, 6 October). Timeline of an uprising, *BBC News*. Retrieved from http://news.bbc.co.uk/1/hi/world/europe/959077.stm

BBC. (2003, 24 December). Djindjic 'sniper' remains silent, *BBC News*. Retrieved from http://news.bbc.co.uk/1/hi/world/europe/3346445.stm

BBC. (2007, 1 November). Bosnian PM resigns over reforms, *BBC News*. Retrieved from http://news.bbc.co.uk/1/hi/world/europe/7072908.stm

BBC. (2008a, 16 September). Dutch block EU-Serbia trade deal, *BBC News*. Retrieved from http://news.bbc.co.uk/1/hi/world/europe/7618275.stm

BBC. (2008b, 1 February). Serbian presidential rivals, *BBC News*. Retrieved from http://news.bbc.co.uk/1/hi/world/europe/7219946.stm

BBC. (2011a, 26 May). Ratko Mladic arrested: Bosnia war crimes suspect held, *BBC News*. Retrieved from http://www.bbc.co.uk/news/world-europe-13561407

BBC. (2011b, 20 July). Serbia holds Croatia war crimes suspect Goran Hadzic, *BBC News*. Retrieved from http://www.bbc.co.uk/news/world-europe-14214573

BBC. (2013, 25 April). Serbian president apologises for Srebrenica 'crime', *BBC News*. Retrieved from http://www.bbc.co.uk/news/world-europe-22297089

Bechev, D., & Andreev, S. (2005). Top-down vs. bottom-up aspects of the EU institution-building strategies in the Western Balkans, Occasional Paper No. 3/05. In S. E. E. S. P. (SEESP) (ed). Oxford: European Studies Centre.

Bell, R. G. (2004). Further up the learning curve: NGOs from transition to Brussels. *Environmental Politics*, 13(1), 194–215.

Beta. (2007, 5 December). Bosnia initials SAA, *B92*. Retrieved from http://www.b92.net/eng/news/region.php?yyyy=2007&mm=12&dd=05&nav_id=45944

Beta. (2009, 21 October). Bosnian talks still deadlocked, *B92*. Retrieved from http://www.b92.net/eng/news/region.php?yyyy=2009&mm=10&dd=21&nav_id=62499

Beta. (2010, 5 May). '720.000 people below poverty line', *B92*. Retrieved from http://www.b92.net/eng/news/society-article.php?yyyy=2010&mm=05&dd=05&nav_id=66924

Beta, & BIRN. (2014, 2 April). Serbian province in line for reduced powers, *Balkan Insight*. Retrieved from http://www.balkaninsight.com/en/article/new-draft-statute-leaves-vojvodina-without-govt-and-capital

Beta & Tanjug. (2013, 10 April). Former minister indicted in corruption case, *B92*. Retrieved from http://www.b92.net/eng/news/crimes.php?yyyy=2013&mm=04&dd=10&nav_id=85638

Bieber, F. (2003). The Serbian opposition and civil society: roots of the delayed transition in Serbia. *International Journal of Politics, Culture, and Society*, 17(1), 73–90.

Bieber, F. (2004). Towards Better Governance with More Complexity? In C. Solioz & T. K. Vogel (eds), *Dayton and Beyond: Perspectives on the Future of Bosnia and Herzegovina* (pp. 74–87). Baden-Baden: Nomos.

BIRN. (2007a, 12 October). Bosnia police reforms: another deadline missed, *B92*. Retrieved from http://www.b92.net/eng/news/region.php?yyyy=2007&mm=10&dd=12&nav_id=44514

BIRN. (2007b, 16 October). EU repeats warning to Bosnia on police reform, *B92*. Retrieved from http://www.b92.net/eng/news/region.php?yyyy=2007&mm=10&dd=16&nav_id=44611

BIRN. (2013, 3 July). EBRD to approve loan for Belgrade Metro, *Balkan Insight*. Retrieved from http://www.balkaninsight.com/en/article/ebrd-to-approve-loan-for-belgrade-metro

Bojicic-Dzelilovic, V., & Kostovicova, D. (2013). Europeanisation and conflict networks: private sector development in post-conflict Bosnia-Herzegovina. *East European Politics, 29*(2), 19–35.

Bomberg, E. (2007). Policy learning in an enlarged European Union: environmental NGOs and new policy instruments. *Journal of European Public Policy, 14*(2), 248–268. doi: 10.1080/13501760601122522

Bomberg, E. & Peterson, J. (2000). Policy transfer and Europeanization: passing the Heineken test? *Queen's Papers on Europeanization*, No. 2/2000. Belfast: Institute of European Studies, Queen's University. Retrieved from https://www.qub.ac.uk/schools/SchoolofPoliticsInternationalStudiesandPhilosophy/FileStore/EuropeanisationFiles/Filetoupload,38445,en.pdf

Börzel, T. A. (2003). *Environmental Leaders and Laggards in Europe: Why There Is (Not) a 'Southern Problem'*. Aldershot: Ashgate.

Börzel, T. A. (2005). Europeanization: How the European Union Interacts with Its Member States. In S. Bulmer & C. Lequesne (eds), *The Member States of the European Union* (pp. 45–76). Oxford: Oxford University Press.

Börzel, T. A. (2007). State capacity and the emergence of new modes of governance. *New Modes of Governance (NEWGOV) project*, Deliverable 12/D8. Berlin: Free University of Berlin.

Börzel, T. A. (2009a). *Coping with Accession to the European Union: New Modes of Environmental Governance*. Basingstoke: Palgrave Macmillan.

Börzel, T. A. (2009b). Environmental Policy and the Challenge of Accession. In T. A. Börzel (ed.), *Coping with Accession to the European Union: New Modes of Environmental Governance* (pp. 32–47). Basingstoke: Palgrave Macmillan.

Börzel, T. A. (2009c). New Modes of Governance and Accession: The Paradox of Double Weakness. In T. A. Börzel (ed.), *Coping with Accession to the European Union: New Modes of Environmental Governance* (pp. 7–31). Basingstoke: Palgrave Macmillan.

Börzel, T. A. (2010). European governance: negotiation and competition in the shadow of hierarchy. *JCMS: Journal of Common Market Studies, 48*(2), 191–219. doi: 10.1111/j.1468-5965.2009.02049.x

Börzel, T. A., & Buzogány, A. (2010a). Governing EU accession in transition countries: the role of non-state actors. *Acta Politica, 45*, 158–182. doi: 10.1057/ap.2009.26

Börzel, T. A., & Buzogány, A. (2010b). Environmental organisations and the Europeanisation of public policy in Central and Eastern Europe: the case of biodiversity governance. *Environmental Politics, 19*(5), 708–735. doi: 10.1080/09644016.2010.508302

Börzel, T. A., & Risse, T. (2005). Public-Private Partnerships. Effective and Legitimate Tools of Transnational Governance? In E. Grande & L. W. Pauly (eds), *Complex Sovereignty: On the Reconstitution of Political Authority in the 21st Century* (pp. 195–216). Toronto: University of Toronto Press.

Bruszt, L., & Vedres, B. (2009). *Fostering Developmental Agency from Without*. Paper presented at the EUSA Eleventh Biennial International Conference, Los Angeles.

Buzogány, A. (2009). Hungary: The Tricky Path of Building Environmental Governance. In T. A. Börzel (ed.), *Coping with Accession to the European Union: New Modes of Environmental Governance* (pp. 123–147). Basingstoke: Palgrave Macmillan.
Carmin, J. (2010). NGO capacity and environmental governance in Central and Eastern Europe. *Acta Politica, 45*(1/2), 183–202.
Carmin, J., & VanDeveer, S. D. (2004). Enlarging EU environments: Central and Eastern Europe from transition to accession. *Environmental Politics, 13*(1), 3–24. doi: 10.1080/09644010410001685119
CEE Bankwatch. (2009a). *Corridor Vc Motorway, Bosnia-Herzegovina*. Paper presented at the EBRD Annual Meeting, London. Retrieved from http://bankwatch.org/documents/VcMotorway_IssuePaper_EBRD_AGM2009.pdf
CEE Bankwatch. (2009b). Report from CEE Bankwatch, Center for Environment and Ekotim Fact-finding Mission on Corridor Vc Motorway Project, Bosnia and Herzegovina, 14–15 September 2009.
CEE Bankwatch. (2010). Report from Fact-finding Mission on Corridor Vc Motorway, Bosnia and Herzegovina, 23–25 February 2010.
CEE Bankwatch. (2012). Gazela Reloaded: Another Illegal Roma Resettlement in Serbia on Account of an EIB Funded Project [Press release]. Retrieved from http://bankwatch.org/news-media/for-journalists/press-releases/gazela-reloaded-another-illegal-roma-resettlement-serbia-a
CEE Bankwatch. (2014, 14 May). How long till the next protests in Bosnia and Herzegovina?, *Bankwatch Mail*. Retrieved from http://bankwatch.org/bwmail/59/how-long-till-next-protests-bosnia-and-herzegovina
CEKOR. (2007). *The Belgrade Bypass and Gazela Bridge Projects*. Paper presented at the EBRD Annual Meeting, Kazan, Russian Federation. http://bankwatch.org/documents/Belgrade_IP.pdf
CEKOR. (2008). Briefing Paper about the 'Gazela' Resettlement Process [Press release]. Retrieved from http://bankwatch.org/sites/default/files/GazelaBriefing_20Oct2008.pdf
Chandler, D. (1999). *Bosnia: Faking Democracy after Dayton*. London; Sterling, VA: Pluto Press.
Chandler, D. (2005). From Dayton to Europe. *International Peacekeeping, 12*(3), 336–349.
Chandler, D. (2006). State-building in Bosnia: the limits of 'informal trusteeship'. *International Journal of Peace Studies, 11*(1).
Clarke, R. (2001). Yugoslavia. In F. W. Carter & D. Turnock (eds), *Environmental Problems of East Central Europe* (pp. 396–416). London: Routledge.
Collantes Celador, G. (2009). Becoming 'European' through police reform: a successful strategy in Bosnia and Herzegovina? *Crime, Law and Social Change, 51*(2), 231–242.
Council of the EU. (2004a). Council Decision of 14 June 2004 on the principles, priorities and conditions contained in the European Partnership with Bosnia and Herzegovina (2004/515/EC).
Council of the EU. (2004b). Council decision of 14 June 2004 on the principles, priorities and conditions contained in the European Partnership with Serbia and Montenegro including Kosovo as defined by the United Nations Security Council Resolution 1244 of 10 June 1999 (2004/520/EC).

Bibliography

Council of the EU. (2006a). Council Decision of 30 January 2006 on the principles, priorities and conditions contained in the European Partnership with Bosnia and Herzegovina and repealing Decision 2004/515/EC (2006/55/EC).

Council of the EU. (2006b). Council decision of 30 January 2006 on the principles, priorities and conditions contained in the European Partnership with Serbia and Montenegro including Kosovo as defined by the United Nations Security Council Resolution 1244 of 10 June 1999 and repealing Decision 2004/520/EC (2006/56/EC).

Council of the EU. (2008a). Council Decision of 18 February 2008 on the principles, priorities and conditions contained in the European Partnership with Bosnia and Herzegovina and repealing Decision 2006/55/EC (2008/211/EC).

Council of the EU. (2008b). Stabilisation and Association Agreement between the European Communities and their Member States, of the one part, and Bosnia and Herzegovina, of the other part – 2008/0073 (AVC): The Council of the European Union.

De Launey, G. (2014). Serbia transforming from pariah to EU partner, *BBC News*. Retrieved from http://www.bbc.co.uk/news/world-europe-25808463

Delegation of the EU to the Republic of Serbia. (2014). A long journey for Serbia to reach European ecology standards.

Dimitrova, A. (2002). Enlargement, institution-building and the EU's administrative capacity requirement. *West European Politics, 25*(4), 171–190.

Dizdarević, S. (2004). The Unfinished State? In C. Solioz & T. K. Vogel (eds), *Dayton and Beyond: Perspectives on the Future of Bosnia and Herzegovina* (pp. 37–44). Baden-Baden: Nomos.

Đokić, I., & Sumpor, M. (2013). The Role of Croatian Civil Society Organisations in the European Union Accession Process. Zagreb: TACSO Croatia.

EC. (2007a). Bosnia and Herzegovina 2007 Progress Report.

EC. (2007b). Serbia 2007 Progress Report.

EC. (2008a). Bosnia and Herzegovina 2008 Progress Report.

EC. (2008b). Serbia 2008 Progress Report.

EC. (2009a). Bosnia and Herzegovina 2009 Progress Report.

EC. (2009b). Serbia 2009 Progress Report.

EC. (2013a). Bosnia and Herzegovina 2013 Progress Report.

EC. (2013b). Serbia 2013 Progress Report.

EIB. (2010). Conclusions Report: Complaint SG/E/2009/07, Gazela Bridge Rehabilitation Project, Belgrade, Serbia, 14 July 2010.

Eising, R. & Koller-Koch, B. (1999). Governance in the European Union: a comparative assessment. In B. Koller-Koch & R. Eising (eds), *The Transformation of Governance in the European Union* (pp. 267–285). London: Routledge.

Elbasani, A. (2013). Europeanization Travels to the Western Balkans: Enlargement Strategy, Domestic Obstacles and Diverging Reforms. In A. Elbasani (ed.), *European Integration and Transformation in the Western Balkans. Europeanization or Business as Usual?* (pp. 3–22). Abingdon: Routledge.

European Commission. (2008). Ad Hoc Evaluation of the CARDS Programme, Boznia-Herzegovina for Democratic Stabilisation, Good Governance, and Economic and Social Development Sectors.

Evans, P. (1995). *Embedded Autonomy: States and Industrial Transformation*. Princeton, NJ: Princeton University Press.

Fagan, A. (1994). Environment and transition in the Czech Republic. *Environmental Politics, 3*(3), 479–494.
Fagan, A. (2006). Neither 'north' nor 'south': the environment and civil society in post-conflict Bosnia-Herzegovina. *Environmental Politics, 15*(5), 787–802. doi: 10.1080/09644010600937215
Fagan, A. (2008). Global-local linkage in the Western Balkans: the politics of environmental capacity building in Bosnia-Herzegovina. *Political Studies, 56*(3), 629–652. doi: 10.1111/J.1467-9248.2007.00711.X
Fagan, A. (2010). *Europe's Balkan Dilemma: Paths to Civil Society or State Building*. London: I.B. Tauris.
Fagan, A., & Jehlicka, P. (2003). Contours of the Czech environmental movement: A comparative analysis of Hnuti duha (Rainbow movement) and Jihoceske matky (South Bohemian mothers). *Environmental Politics, 12*(2), 49–70.
Fagan, A., & Sircar, I. (2010a). Compliance without governance: the role of NGOs in environmental impact assessment processes in Bosnia-Herzegovina. *Environmental Politics, 19*(4), 599–616. Pii 925519691, doi: 10.1080/09644016.2010.489714
Fagan, A., & Sircar, I. (2010b). Environmental politics in the Western Balkans: river basin management and non-governmental organisation (NGO) activity in Herzegovina. *Environmental Politics, 19*(5), 808–830.
Fagan, A., & Sircar, I. (2011). Foreign donor assistance and environmental capacity building: evidence from Serbia and Bosnia-Herzegovina. *European Political Science Review, 3*(2), 301–320. doi: 10.1017/S1755773910000391
Falkner, G., & Treib, O. (2008). Three worlds of compliance or four? The EU 15 compared to new member states. *JCMS: Journal of Common Market Studies, 46*(2), 293–313.
Figa, J. (1997). Socializing the State: Civil Society and Democratization from Below in Slovenia. In M. K. Bokovoy, J. A. Irvine, & C. S. Lilly (eds), *State-Society Relations in Yugoslavia 1945–1992* (pp. 163–182). Basingstoke: Palgrave Macmillan.
Frandino, N. (2014, 10 February). Solidarity protests for Bosnia reach neighboring Serbia [video], *Reuters*. Retrieved from http://uk.reuters.com/video/2014/02/10/solidarity-protests-for-bosnia-reach-nei?videoId=276800719
Gardner, A. (2013, 4 December). Tentative agreements reached between Bosnian political leaders, *European Voice*. Retrieved from http://www.europeanvoice.com/article/tentative-agreements-reached-between-bosnian-political-leaders/
Gec, J. (2012, 21 May). Tomislav Nikolic, 'Toma the Gravedigger,' elected Serbian president, *Huffington Post*. Retrieved from http://www.huffingtonpost.com/2012/05/21/tomislav-nikolic-serbia-president_n_1532736.html
Geoghegan, P. (2014, 14 May). Welcome to Brčko, Europe's only free city and a law unto itself, *Guardian (UK)*. Retrieved from http://www.theguardian.com/cities/2014/may/14/brcko-bosnia-europe-only-free-city
George, A. L., & Bennett, A. (2005). *Case Studies and Theory Development in the Social Sciences*. Cambridge, Mass: MIT Press.
Goetz, K. H. (2008). Governance as a path to government. *West European Politics, 31*(1–2), 258–279. doi: 10.1080/01402380701835066
Goetz, K. H., & Wollmann, H. (2001). Governmentalizing central executives in post-communist Europe: a four-country comparison. *Journal of European Public*

Policy, 8(6), 864–887. Government of the Republic of Serbia. (2010). Government to Finance Horgos-Novi Sad Motorway Section [Press release]

Grabbe, H. (1999). A Partnership for Accession? The Implications of EU Conditionality for the Central and East European Applicants. *Robert Schuman Centre Working Paper*. San Domenico di Fiesole (FI): European University Institute.

Grabbe, H. (2001). How does Europeanization affect CEE governance? Conditionality, diffusion and diversity? *Journal of European Public Policy, 8*(6), 1013–1031.

Grabbe, H. (2006). *The EU's Transformative Power: Europeanization through Conditionality in Central and Eastern Europe*. Basingstoke: Palgrave Macmillan.

Grindle, M. S. (2004). Good enough governance: poverty reduction and reform in developing countries. *Governance: An International Journal of Policy and Administration, 17*(4), 525–548.

Guttenbrunner, S. (2009). Poland: When Environmental Governance Meets Politics. In T. A. Börzel (ed.), *Coping with Accession to the European Union: New Modes of Environmental Governance* (pp. 148–168). Basingstoke: Palgrave Macmillan.

Hallstrom, L. K. (2004). Eurocratising enlargement? EU elites and NGO participation in European environmental policy. *Environmental Politics, 13*(1), 175–193.

Hellman, J. S., Jones, G., & Kaufmann, D. (2000). '*Seize the State, Seize the Day*': *State Capture, Corruption, and Influence in Transition*. Policy Research Working Paper 2444. World Bank, World Bank Institute. Governance, Regulation, and Finance Division, & Europe and Central Asia Region, Public Sector Group, & European Bank for Reconstruction and Development, Office of the Chief Economist. Retrieved from http://elibrary.worldbank.org/doi/pdf/10.1596/1813-9450-2444

Héritier, A. (2003). Composite democracy in Europe: the role of transparency and access to information. *Journal of European Public Policy, 10*(5), 814–833. doi: 10.1080/1350176032000124104

Héritier, A. (2005). Europeanization Research East and West: A Comparative Assessment. In F. Schimmelfennig & U. Sedelmeier (eds), *The Europeanization of Central and Eastern Europe* (pp. 199–209). Ithaca, NY: Cornell University Press.

Héritier, A., & Lehmkuhl, D. (2008). The shadow of hierarchy and new modes of governance. *Journal of Public Policy, 28*(1), 1–17.

Hicks, B. (2004). Setting Agendas and Shaping Activism: EU Influence on Central and Eastern European Environmental Movements. *Environmental Politics, 13*(1), 216–233.

Howard, M. M. (2003). *The Weakness of Civil Society in Post-Communist Europe*. Cambridge; New York: Cambridge University Press.

Hughes, J., Sasse, G., & Gordon, C. E. (2004). *Europeanization and Regionalization in the EU's Enlargement to Central and Eastern Europe: The Myth of Conditionality*. Basingstoke; New York: Palgrave Macmillan.

Hyde-Price, A. G. V. (1996). *The International Politics of East Central Europe*. Manchester; New York: Manchester University Press: St. Martin's Press [distributor].

Illmer, A. (2011, 5 February). Serbian opposition rally calls for early elections, *Deutsche Welle*. Retrieved from http://www.dw.de/serbian-opposition-rally-calls-for-early-elections/a-14820181

IMF. (2004). Bosnia and Herzegovina: Poverty Reduction Strategy Paper, Mid-Term Development Strategy. Washington, DC: IMF.
Jacoby, W. (2004). *The Enlargement of the EU and Nato: Ordering from the Menu in Central Europe*. Cambridge: Cambridge University Press.
Jancar-Webster, B. (1998). Environmental movement and social change in the transition countries. *Environmental Politics, 7*(1), 69–90.
Jehlicka, P., & Tickle, A. (2004). Environmental implications of eastern enlargement: the end of progressive EU environmental policy? *Environmental Politics, 13*(1), 77–95.
Jervis, R. (2006). Understanding beliefs. *Political Psychology, 27*(5), 641–663.
Jordan, A. (1999). The implementation of EU environmental policy: a policy problem without a political solution? *Environment and Planning C: Government and Policy, 17*(1), 69–90.
Joseph, E. P., & Hitchner, R. B. (2008, June). Making Bosnia Work: Why EU Accession Is Not Enough. *The United States of Peace*. Retrieved from http://www.usip.org/publications/making-bosnia-work-why-eu-accession-not-enough
Juncos, A. E. (2005). The EU's post-conflict intervention in Bosnia and Herzegovina: (re)integrating the Balkans and/or (re)inventing the EU? *Southeast European Politics, 6*(2), 88–108.
Juncos, A. E. (2012). Member state-building versus peacebuilding: the contradictions of EU state-building in Bosnia and Herzegovina. *East European Politics, 28*(1), 58–75.
Keil, S. (2014, 16 June). Whatever happened to the Plenums in Bosnia?, *Balkan Insight*. Retrieved from http://www.balkaninsight.com/en/article/whatever-happened-to-the-plenums-in-bosnia
Kelava, M. (2012, 6 June). Devet dana banjalučke "varšavske" + FOTOGALERIJA + VIDEO!, *H-alter*. Retrieved from http://www.h-alter.org/vijesti/devet-dana-banjalucke-varsavske-fotogalerija-video
Kenis, P., & Schneider, V. (1991). Policy Networks and Policy Analysis: Scrutinizing a New Analytical Toolbox. In B. Marin & R. Mayntz (eds), *Policy Networks: Empirical Evidence and Theoretical Considerations* (pp. 25–59). Boulder, CO: Westview Press.
Ker-Lindsay, J. (2010, 22 July). A victory of sorts for Kosovo, *The Guardian*. Retrieved from http://www.theguardian.com/commentisfree/2010/jul/22/victory-of-sorts-kosovo
Knaus, G., & Martin, F. (2003). Travails of the European Raj. *Journal of Democracy, 14*(3), 60–74.
Knill, C., & Lehmkuhl, D. (1999). How Europe Matters. Different Mechanisms of Europeanization. European Integration online Papers (EIoP).
Kohler-Koch, B. (1999). The Evolution and Transformation of European Governance. In B. Kohler-Koch & R. Eising (eds), *The Transformation of Governance in the European Union* (pp. 14–35). London: Routledge.
Kostovicova, D., & Bojicic-Dzelilovic, V. (2008). *Transnationalism in the Balkans*. London: Routledge.
Kostovicova, D., & Bojicic-Dzelilovic, V. (2011). External Statebuilding and Transnational Networks: The Limits of the Civil Society Approach. In D. Kostovicova & M. Glasius (eds), *Bottom-Up Politics: An Agency-Centred Approach to Globalization* (pp. 93–111). Basingstoke: Palgrave Macmillan.

Krunic-Lazic, M. (2010). *Energy and Environment in Serbia*. Paper presented at the EIONET workshop on energy and environment, Copenhagen, Denmark.

Kubicek, P. J. (2003). International Norms, the European Union, and Democratization: Tentative Theory and Evidence. In P. J. Kubicek (ed.), *The European Union and Democratization* (pp. 1–29). London: Routledge.

Kutter, A., & Trappmann, V. (2010). Civil society in Central and Eastern Europe: the ambivalent legacy of accession. *Acta Politica, 45*(1/2), 41–69.

Lemos, M. C., & Agrawal, A. (2006). Environmental governance. *Annual Review of Environment and Resources, 31*, 297–325.

Lomax, B. (1997). The strange death of 'civil society' in post-communist Hungary. *Journal of Communist Studies and Transition Politics, 13*(1), 41–63.

Lungescu, O. (2010, 14 June). EU unfreezes Serbia trade accord, *BBC News*. Retrieved from http://www.bbc.co.uk/news/10314765

Lynch, L. (2012, 18 July). Serbian court disputes limits of Vojvodina autonomy, *Southeast European Times*. Retrieved from http://www.setimes.com/cocoon/setimes/xhtml/en_GB/features/setimes/features/2012/07/18/feature-01

Majone, G. (1996). *Regulating Europe*. London; New York: Routledge.

Mandel, R. (2002). Seeding civil society. In C. M. Hann (ed.), *Postsocialism: Ideals, Ideologies, and Practices in Eurasia* (pp. 279–296). London; New York: Routledge.

Manners, I. (2002). Normative power Europe: a contradiction in terms? *Journal of Common Market Studies, 40*(2), 235–258.

March, J. G., & Olsen, J. P. (1989). *Rediscovering Institutions: The Organizational Basis of Politics*. New York: Palgrave Macmillan.

March, J. G., & Olsen, J. P. (2006). The logic of appropriateness. In M. Moran, M. Rein, & R. E. Goodin (eds), *The Oxford Handbook of Public Policy* (pp. 689–708). Oxford: Oxford University Press.

Mayntz, R. (2003). New Challenges to Governance Theory. In H. P. Bang (ed.), *Governance as Social and Political Communication* (pp. 27–40). Manchester: Manchester University Press.

Mayntz, R. & Scharpf, F. W. (1995). Steuerung und Selbstorganisation in staatsnahen Sektoren. In R. Mayntz & F.W. Scharpf (eds), *Gesellschaftliche Selbstregelung und politische Steuerung* (pp. 9–38). Frankfurt am Main: Campus.

Migdal, J. S. (1988). *Strong Societies and Weak States: State-Society Relations and State Capabilities in the Third World*. Princeton, NJ: Princeton University Press.

Miller, N. J. (1997). Reconstituting Serbia: 1945–1991. In M. K. Bokovoy, J. A. Irvine, & C. S. Lilly (eds), *State-Society Relations in Yugoslavia, 1945–1992* (pp. 291–314). New York: St. Martin's Press.

Ministry of Environment of Serbia. (2009). The National Assembly Adopted 16 Laws in the Field of Environment [Press release].

Muehlmann, T. (2008). Police restructuring in Bosnia-Herzegovina: problems of internationally-led security sector reform. *Journal of Intervention and Statebuilding, 2*(1), 1–22.

Nowak, M. (2004). Has Dayton Failed? In C. Solioz & T. K. Vogel (eds), *Dayton and Beyond: Perspectives on the Future of Bosnia and Herzegovina* (pp. 45–58). Baden-Baden: Nomos.

Obradovic, D., & Pleines, H. (eds). (2007). *The Capacity of Central and East European Interest Groups to Participate in EU Governance*. Stuttgart: Ibidem-Verlag.

Ostojić, M. (2014). *Between Justice and Stability: The Politics of War Crimes Prosecutions in Post-Milošević Serbia*. Farnham: Ashgate.
Parau, C. E. (2009). Impaling Dracula: how EU accession empowered civil society in Romania. *West European Politics, 32*(1), 119–141. doi: 10.1080/01402380802509917; Pii 906340758
Parau, C. E. (2010). East side story: how transnational coalitions contested EU conditionality. *Europe-Asia Studies, 62*(9), 1527–1554.
Partos, G. (2003, 12 March). Obituary: Zoran Djindjic, *BBC News*. Retrieved from http://news.bbc.co.uk/1/hi/world/europe/2843451.stm
Petrova, T., & Tarrow, S. (2007). Transactional and participatory activism in the emerging European polity the puzzle of East-Central Europe. *Comparative Political Studies, 40*(1), 74–94. doi: 10.1177/0010414006291189
Phinnemore, D. (2013). The Stabilization and Association Process: A Framework for European Union Enlargement? In A. Elbasani (ed.), *European Integration and Transformation in the Western Balkans: Europeanization or business as usual?* (pp. 22–35). Abingdon: Routledge.
Pierre, J. (2000). *Debating Governance*. Oxford; New York: Oxford University Press.
Pop-Eleches, G. (2007). Historical legacies and post-communist regime change. *Journal of Politics, 69*(4), 908–926.
Pridham, G. (2000). *The Dynamics of Democratization: A Comparative Approach*. New York: Continuum.
Radaelli, C. M. (2000). *Whither Europeanization? Concept Stretching and Substantive Change*. European Integration online Papers (EIoP).
Radaelli, C. M. (2003). The Europeanization of Public Policy. In K. Featherstone & C. M. Radaelli (eds), *The Politics of Europeanization* (pp. 27–56). Oxford: Oxford University Press.
Radio Slobodna Evropa. (2012, 23 June). Buđenje se dogodilo i ništa ga ne može zaustaviti, *Radio Slobodna Evropa*. Retrieved from http://www.slobodnaevropa.org/content/budjenje-se-dogodilo-i-nista-ga-ne-moze-zaustaviti/24623653.html
Raik, K. (2004). EU accession of Central and Eastern European countries: democracy and integration as conflicting logics. *East European Politics and Societies, 18*(4), 567–594.
REC. (2006). *NGO Directory of South Eastern Europe: A Directory and Survey Findings of West Balkan Environmental Civil Society Organisations*. Szentendre: Regional Environmental Center.
Rhodes, R. A. W. (1997). *Understanding Governance: Policy Networks, Governance, Reflexivity and Accountability*. Buckingham: Open University Press.
Rootes, C. A. (2003). *Environmental Protest in Western Europe*. Oxford: Oxford University Press.
Rosenau, J. N., & Czempiel, E. O. (1992). *Governance without Government: Order and Change in World Politics*. Cambridge; New York: Cambridge University Press.
Rowland, J. (2000, 24 March). Milosevic still standing strong, *BBC News*. Retrieved from http://news.bbc.co.uk/1/hi/world/europe/688747.stm
Sbragia, A. (2000). Environmental Policy: Economic Constraints and External Pressures. In H. Wallace & W. Wallace (eds), *Policy-making in the European Union*, 4th ed. (pp. 293–316). Oxford: Oxford University Press.

Scharpf, F. W. (1978). Interorganizational Policy Studies: Issues, Concepts and Perspectives. In K. Hanf & F. W. Scharpf (eds), *Interorganizational Policy Making. Limits to Coordination and Central Control* (pp. 57–112). London: Sage.

Schimmelfennig, F., & Sedelmeier, U. (2004). Governance by conditionality: EU rule transfer to the candidate countries of Central and Eastern Europe. *Journal of European Public Policy, 11*(4), 661–679. doi: 10.1080/1350176042000248089

Schimmelfennig, F., & Sedelmeier, U. (2005a). Conclusions: The Impact of the EU on the Accession Countries. In F. Schimmelfennig & U. Sedelmeier (eds), *The Europeanization of Central and Eastern Europe* (pp. 201–228). Ithaca, NY: Cornell University Press.

Schimmelfennig, F., & Sedelmeier, U. (2005b). *The Europeanization of Central and Eastern Europe*. Ithaca, NY: Cornell University Press.

Schimmelfennig, F., & Sedelmeier, U. (2005c). Introduction: Conceptualizing the Europeanization of Central and Eastern Europe. In F. Schimmelfennig & U. Sedelmeier (eds), *The Europeanization of Central and Eastern Europe* (pp. 1–28). Ithaca, NY: Cornell University Press.

Sedelmeier, U. (2011). Europeanisation in new member and candidate states. *Living Reviews in European Governance, 6*(1).

SEIO. (2013). National Plan for the Adoption of the Acquis (2013–2016). Belgrade: Serbian European Integration Office (SEIO). Retrieved from http://www.seio.gov.rs/upload/documents/nacionalna_dokumenta/npaa13_16.pdf

Sherlock, K. L., Kirk, E. A., & Reeves, A. D. (2004). Just the usual suspects? Partnerships and environmental regulation. *Environment and Planning C: Government and Policy, 22*(5), 651–666.

Simon, H. A. (1985). Human nature in politics: the dialogue of psychology with political science. *American Political Science Review, 79*(2), 293–304.

Sissenich, B. (2007). *Building States without Society: European Union Enlargement and the Transfer of EU Social Policy to Poland and Hungary*. Lanham, MD: Lexington Books.

Sissenich, B. (2010). Weak states, weak societies: Europe's east-west gap. *Acta Politica, 45*(1/2), 11–40.

Statistical Office of the Republic of Serbia. (2009). Municipalities of Serbia 2008.

Stuppert, W. (2010). Sustainability of Civil Society Organizations in Serbia. Belgrade: OSCE Mission to Serbia. Retrieved from http://www.osce.org/serbia/68666

Sudbery, I. (2010). The European Union as political resource: NGOs as change agents? *Acta Politica, 45*(1/2), 136–157.

Szulecka, J., & Szulecki, K. (2013). Analysing the Rospuda River controversy in Poland: rhetoric, environmental activism, and the influence of the European Union. *East European Politics, 29*(4), 397–419.

Tanjug. (2008a, 17 June). Bosnia's European perspective confirmed, *B92*. Retrieved from http://www.b92.net/eng/news/region.php?yyyy=2008&mm=06&dd=17&nav_id=51151

Tanjug. (2008b, 30 April). 'DS experts at dirty signatures', *B92*. Retrieved from http://www.b92.net/eng/news/politics.php?yyyy=2008&mm=04&dd=30&nav_id=49868

Tanjug. (2009a, 22 December). Serbia submits formal EU candidacy bid, *B92*. Retrieved from http://www.b92.net/eng/news/politics.php?yyyy=2009&mm=12&dd=22&nav_id=63937

Tanjug. (2009b, 24 December). Statute motion filed with Constitutional Court, *B92*. Retrieved from http://www.b92.net/eng/news/politics.php?yyyy=2009&mm=12&dd=24&nav_id=64003

Tanjug. (2010a, 14 June). European Union approves SAA ratification, *B92*. Retrieved from http://www.b92.net/eng/news/politics.php?yyyy=2010&mm=06&dd=14&nav_id=67784

Tanjug. (2010b, 2 January). Pajtić expects statute to go into effect, *B92*. Retrieved from http://www.b92.net/eng/news/politics.php?yyyy=2010&mm=01&dd=02&nav_id=64195

Tanjug. (2013a, 19 April). Belgrade, Priština initial agreement in Brussels, *B92*. Retrieved from http://www.b92.net/eng/news/politics.php?yyyy=2013&mm=04&dd=19&nav_id=85780

Tanjug. (2013b, 22 April). Government adopts Brussels agreement, *B92*. Retrieved from http://www.b92.net/eng/news/politics.php?yyyy=2013&mm=04&dd=22&nav_id=85829

Tanjug. (2013c, 9 March). Serbian reps inform EC about reform progress, *B92*. Retrieved from http://www.b92.net/eng/news/politics.php?yyyy=2013&mm=03&dd=09&nav_id=85080

Tanjug. (2014a, 22 May). Assembly of Vojvodina adopts provincial statute, *Tanjug*. Retrieved from http://www.tanjug.rs/news/130179/assembly-of-vojvodina-adopts-provincial-statute.htm

Tanjug. (2014b, 25 March). Commission announces final election results, *B92*. Retrieved from http://www.b92.net/eng/news/politics.php?yyyy=2014&mm=03&dd=25&nav_id=89760

Tanjug. (2014c, 27 April). Parliament elects new Serbian government, *B92*. Retrieved from http://www.b92.net/eng/news/politics.php?yyyy=2014&mm=04&dd=27&nav_id=90121

Tanjug. (2014d, 16 September). Screening starts for Chapter 27 – Environment, *Tanjug*. Retrieved from http://www.tanjug.rs/news/145115/screening-starts-for-chapter-27---environment.htm

Tanjug, & DPA. (2007, 30 November). Political crisis in Bosnia resolved, *B92*. Retrieved from http://www.b92.net/eng/news/region.php?yyyy=2007&mm=11&dd=30&nav_id=45815

The Government of the Republic of Serbia. (2009). *Amended National Programme for Integration of the Republic Of Serbia into the European Union (Abridged Version)*. Belgrade.

Thorpe, N. (2008, 11 September). Serbs' EU hopes still hang on The Hague, *BBC News*. Retrieved from http://news.bbc.co.uk/1/hi/world/europe/7611150.stm

Traynor, I. (2012, 20 May). Former extreme nationalist becomes Serbian president, *The Guardian*. Retrieved from http://www.theguardian.com/world/2012/may/20/serbia-president-eu

UNEP, and UNCHS. (1999). *The Kosovo Conflict: consequences for the environment and human settlements*. Retrieved from http://www.grid.unep.ch/btf/final/finalreport.pdf

Vachudova, M. A. (2005). *Europe Undivided: Democracy, Leverage, and Integration after Communism*. Oxford: Oxford University Press.

VanDeveer, S. D., & Carmin, J. (2005). EU Environmental Policy and the Challenges of Eastern Enlargement. In A. Jordan (ed.), *Environmental Policy* in the European Union: Actors, Institutions, and Processes 2nd ed. (pp. 279–294). London: Earthscan.

Večernje novosti. (2008, 8 October). Peti park pobedio!, *Večernje novosti*. Retrieved from http://www.novosti.rs/vesti/beograd.74.html:224340-Peti-park-pobedio

Večernje novosti. (2010, 16 June). '3 EU states against Serbia's bid', *B92*. Retrieved from http://www.b92.net/eng/news/politics.php?yyyy=2010&mm=06&dd=16&nav_id=67845

Vennesson, P. (2008). Case Studies and Process Tracing: Theories and Practices. In D. Della Porta & M. Keating (eds), *Approaches and Methodologies in the Social Sciences: A Pluralist Perspective* (pp. 223–239). Cambridge: Cambridge University Press.

Vogel, D., & Kagan, R. A. (2004). National Regulations in a Global Economy. In D. Vogel & R. A. Kagan (eds), *Dynamics of Regulatory Change: How Globalization Affects National Regulatory Policies* (pp. 1–41). London: University of California Press.

Waller, M. (2010). The environmental issue in the east of Europe: top-down, bottom-up and outside-in. *Environmental Politics, 19*(5), 831–849.

Weiss, L. (1998). *The Myth of the Powerless State*. Ithaca, NY: Cornell University Press.

Index

Aarhus Convention, 36, 44, 63, 106, 126, 144, 170*n*22
acquis communautaire, 15, 99, 148
Ahtisaari Plan, 109
ANPED (Northern Alliance for Sustainability), 123, 165*n*47
Ashton, Catherine, 102

Bajina Bašta, 113, 115, 116, 172*n*53, 172*n*61
Banja Luka, 53, 54, 57, 83–5, 87, 122, 157, 160
Banja Luka–Gradiška Road, 66–7, 73–5, 79–81, 84, 166*n*20–1
Barcelona Convention, 44
BCEOM (French company), 73, 81
BCIF (Balkans Community Initiatives Fund), 121, 123, 174*n*106
Bela breza, 123, 124, 174*n*106
Belgrade, 8, 81, 93, 96, 97, 108, 119, 121–3, 125, 160, 161
 federal authorities, 134–6
 municipalities, 110–12, 114–15, 117
 negotiations with Pristina, 101–3, 153
 Zagreb–Belgrade motorway, 67
Belgrade Bypass, 129, 131, 140, 142
Bern Convention, 44
BiH (Bosnia and Herzegovina)
 Banja Luka–Gradiška Road, 66–7
 Buna–Neum Road (Neum–Stolac section), 69–70
 cantonal-level capacities, 49–52
 case studies in, 65–73
 challenges of Europeanizing, 39–42
 conditionality, 36–37
 Corridor 5c between Blagaj and Počitelj, 72–3
 EIAs and environmental governance in, 151–3
 entity-level capacities, 45–9
 environmental governance in, 20–1
 EU building environmental governance, 34–7, 61–2
 Europeanization of, 5–7, 29–31
 financial and technical assistance, 34–6
 Mostar Bypass, 70–1
 multi-level power-sharing agreement, 32–4
 municipal level, 52–5
 non-state actors, 56–60
 political future, 158–62
 Sarajevo Bypass, 67–9
 state-level capacities, 42–5
 weakness of EU conditionality in, 37–9
Bildt, Carl, 39
Blagaj–Počitelj, 66, 142
 Corridor 5c between, 72–3
 mobilization around, 86–8
Bosnia–Herzegovina, *see* BiH (Bosnia and Herzegovina)
Bosnian Podrinje Canton, 51–2
Brammertz, Serge, 100, 103
Brussels, 18–19, 28, 31, 59, 61, 98, 102, 109, 118, 122–3, 157
Brussels Agreement, 153
Bulgaria, 15, 30, 59, 129, 155, 160
Bulgarian border, 129, 141, 144, 146, 174*n*6, 177*n*70, 177*n*73, 177*n*75–6
Buna–Neum Road (Neum–Stolac section), 66, 69–70, 76, 82, 85–6

CARDS (Community Assistance for Reconstruction, Development and Stabilisation), 34–5, 97, 99, 123, 163*n*1
Cartagena Protocol, 44

191

CEE (Central and East European) states, 2, 4, 13
  EU accession process for, 26–8, 30–1
  EU influencing governance, 16
  EU transforming, 41–4
  external governance, 14–15
  externally generated new modes of governance, 23–9
CEE Bankwatch, 72, 78, 87, 122, 142, 143, 157, 161, 176n59
CEKOR (Center for Ecology and Sustainable Development), 87, 122, 131, 142–4,157
CESD (Centre for Sustainable Development), 58, 84
CeSID (Center for Free Elections and Democracy), 93
CIP (Traffic Institute, Belgrade), 139, 140, 176n50, 176n53
CITES (Convention on International Trade in Endangered Species), 44
conditionality
  BiH, 36–7
  EU, 2–5, 13–19, 21, 26, 34, 36–7, 45, 97–9, 147–8, 153, 155, 157, 161
  influence of EBRD on EIA, 80, 89
  limits of EU, in Serbia, 99–103
  long-term partnership, 103–4
  Serbia, 97–9
  weakness of EU, in BiH, 37–9
Corridor 5c, 50, 54, 62, 66, 67–8, 70, 71, 77–80, 84–5, 86–8, 122, 158
Corridor 10, 68, 98, 111, 129, 134–7, 175n23, 175n33, 176n53, 176n55
  Corridor 10b, 129
  Corridor 10c, 129–30
  Corridor 10d, 130
  developers for, 137–9
  EIA processes, 111, 134–7, 141–7, 148
  funding, 132–3
  technical consultants, 139–41
Corridors of Serbia, 8, 146, 175n33
CPV (Centre for Roads of Vojvodina, Novi Sad), 139–40, 148

Croatia, 1, 2, 3, 7, 40, 45, 59–60, 62, 68–70, 92, 94, 109, 148, 160, 171n26, 177n2
Green Action, 165n47
Crveni Krst, 111, 115, 172n61
CSOs (civil society organizations), 26–7, 31, 56–60, 83, 112, 120, 146, 148, 155, 161
Czech Republic, 15, 30, 116, 155
CZZS (Centre for Environment), 57, 83, 87, 122–3, 142, 157, 160, 165n47

Dačić, Ivica, 95, 96
Dayton–Paris Agreement (1995), 3, 32, 33, 34, 35, 40, 41, 166n13
democracy, 1, 2, 97, 99, 155
Department for Environmental Protection, 43, 53, 54, 115
DEI (Directorate of European Integration), 42–3, 48
Djindjic, Zoran, 93, 94
Dodik, Milorad, 38, 163n9
Donji Vakuf, 54, 55, 82, 164n31
DOS (Democratic Opposition of Serbia), 93, 94, 121
DS (Democratic Party), 93–6
DSS (Democratic Party of Serbia), 93–5, 99, 100, 109
Dulić, Oliver, 106, 170n20

EAP (Environmental Action Plan), 66, 67, 80
EAR (European Agency for Reconstruction), 96–7, 112, 116, 118
EBRD (European Bank for Reconstruction and Development), 66, 67, 74, 79–82, 84, 87, 89, 122, 128, 132–3, 137, 139–40, 142–4, 153–4, 161, 166n6, 167n51, 175n10
ECHR (European Court of Human Rights), 39
EEA (European Environmental Agency), 107, 171n26
EEB (European Environmental Bureau), 59, 157

Index 193

EIA (environmental impact assessment) in BiH, 8, 37, 44, 48, 56, 62
  Banja Luka–Gradiška Road, 66–7
  BiH and Serbia, 151–3
  Buna–Neum Road (Neum–Stolac section), 69–70
  case studies in BiH, 65–73
  Corridor 5c between Blagaj and Počitelj, 72–3
  EU Directive on EIA, 65
  implementation of, 88–90
  international actors and social learning, 79–82
  laws, 6–7
  mobilization around Blagaj–Počitelj, 86–8
  Mostar Bypass, 70–1
  municipal governments, 54–5
  NGOs and local communities, 83–6
  Sarajevo Bypass, 67–9
  state and public actors, 73–9
  studies at FMOIT, 46–7
EIA (environmental impact assessment) in Serbia, 126–38, 151–3
  Belgrade Bypass, 131
  case studies, 128–31
  Corridor 10b, 129
  Corridor 10c, 129–30
  Corridor 10d, 130
  developers for Corridor 10, 137–9
  EU EIA Directive Law in Serbia, 128
  governmental structures, 134–7
  IFIs (international financial institutes), 128, 132–3
  NGOs and local communities, 141–7
  technical consultants, 139–41
EIB (European Investment Bank), 66, 67, 79–82, 90, 128, 132, 139, 143–4, 176n59
EIDHR (European Initiative for Democracy and Human Rights), 35, 97, 169n3

EIONET (European environment information and observation network), 107, 171n26
Eko Jasenica, 85
Eko Most, 85
Eko Plan (Mostar), 76, 82
Ekotim, 57, 85, 87, 122, 142, 168n74
EMP (Environmental Management Plan), 80, 132, 133, 138, 141, 145–6, 148, 176n45, 176n55
ENGOs (environmental non-governmental organizations), 31, 56, 58–61, 75, 83–5, 87, 89–90, 146, 156
  in BiH, 142, 148
  Sarajevo Bypass EIA, 84
  in Serbia, 118–24, 125, 142
  transnational European networks, 157
  see also NGOs (non-governmental organizations)
enlargement
  Europeanization via, 10–19
  exceptional case of eastward, 12–15
Environmental Ambassadors for Sustainable Development, 121, 170n19–20
environmental governance
  BiH, 20–1
  BiH and Serbia, 151–3
  development of multi-level, 8
  DEI (Directorate of European Integration), 42–3, 48
  environment and, 4–5
  and EU enlargement, 29–31
  governance to, 19–29
  MOFTER (Ministry for Foreign Trade and Economic Relations), 42–5, 47–8, 63, 76
  state competencies in Serbia, 104–8
environmental regulation, 4, 29, 106, 152, 154, 156
EPD (Enhanced Permanent Dialogue), 98, 170n8
E-PRTR (European Pollutant Release and Transfer Register), 107

EPUS (European Movement in Serbia), 117–18
Espoo Convention, 36, 44, 45, 64, 127
EU (European Union)
  accession strategy for CEE states, 26–8, 30–1
  building environmental governance in BiH, 34–37
  conditionality, 2–5, 13–19, 21, 26, 34, 36–7, 45, 97–9, 147–8, 153, 155, 157, 161
  decision maker and enforcer, 22–3
  engagement in Serbia, 96–9
  environmental governance and EU enlargement, 29–31
  Europeanization and impact of, 153–8
  influencing development of governance, 16–19
  new modes of governance, 21–3
  principal driver of reforms, 16
  record and legacy, 29–31
  weakness of EU conditionality in BiH, 37–9
EU Road Map, 36
European Commission, 34–6, 37–9, 79, 89–90, 101–2, 105, 163n2
European Council, 36, 98, 100
Europeanization
  BiH, 5–7
  case of eastward enlargement, 12–15
  definition of, 11–12
  impact of EU on, 153–8
  new modes of governance, 21–3
  political future of BiH and Serbia, 158–62
  Serbia, 5–7
  variation in pace of reform and compliance, 15–19
  via enlargement, 10–19
European Neighbourhood Policy, 11
European Partnership, 36, 98–9, 155

FBiH (Federation of BiH), 32–3, 63–4, 66, 71, 82, 84, 136, 160
  cantonal-level governance, 49–52
  entity-level governance, 45–8

environmental governance, 44–5
FMOIT (Federal Ministry for Environment and Tourism), 45–7, 50–2, 65, 75–8, 81–2, 85–6, 89, 165n4
  laws on environmental protection, 65, 165n1–4
Ministries for Environmental Protection, 42
municipal-level governance, 52–5, 60
policy co-ordination, 40–2
Roads Directorate, 80, 82, 87, 89, 148
state and public actors, 73–9
FDI (foreign direct investment), 1, 56, 59–60
Federal Roads Directorate, 74, 75, 167n31
financial and technical assistance
  BiH, 34–6
  Serbia, 97
FMOIT (Federal Ministry for Environment and Tourism), 45–7, 50–2, 65, 75–8, 81–2, 85–6, 89, 165n4
Fondeko, 84
FYEG (Federation of Young European Greens), 122

Gazela Bridge, 142–3
governance
  EU influencing development of, 16–19
  externally generated new modes of, 23–9
  new modes of, 21–3
  transition to environmental, 19–29
governance capacity, 23, 66, 118, 126
Greece, 130, 160, 177n1
Green 10 organizations, 59, 90, 121–2, 157
Green Bike, 123, 124, 174n106
Green List of Serbia, 123–4
Green Network of Vojvodina, 111–12, 122, 148
GTZ (German Agency for Technical Co-operation), 116, 118

Index 195

Hadžić, Goran, 101, 102, 103, 153
HEAL (Health and Environment Alliance), 157–8
Herzegovina, *see* BiH (Bosnia and Herzegovina)
Herzogovina-Neretva Canton, 33, 50–2, 71, 76–7, 86–7
HIP (Highway Institute), 132, 133, 137, 139–41, 142, 148
human rights, 34, 35, 36, 97, 119–20, 142, 143
Hungary, 30, 56, 59, 66, 68, 163n1

ICJ (International Court of Justice), 101–2
ICTY (International Criminal Tribunal for Yugoslavia), 3, 34, 61, 92–4, 98–101, 103, 147, 154
IFIs (international financial institutions), Corridor 10 project in Serbia, 128, 132–3
IGH (Civil Engineering Institute of Croatia, Zagreb), 77, 78
Institute for Nature Protection, Serbia, 108, 110, 134, 135, 140, 145
IPA (Instrument for Pre-Accession Assistance), 35, 62, 97, 98, 99, 104, 113, 124, 153–4, 163n1, 163n3
IPSA (Sarajevo-based engineering firm), 77, 87

Josef Pančić, 123, 124, 174n106

K10 (Corridor 10 LLC), 132, 137–9, 141, 142, 143, 146, *see also* Corridors of Serbia
Karadžić, Radovan, 100
Koštunica, Vojislav, 93
Kosovo, 1, 3, 41, 93, 98, 99, 169n2
 conflict, 124
 conflict in, 91–2, 96
 independence, 94–6, 100
 status, 154, 159–60
 status of, 101–2, 108, 109–10
Kumanovo Agreement, 109
Kyoto Protocol, 37, 48, 106, 170n22–3

Lajčák, Miroslav, 38, 40
LEAP (Local Environmental Action Plan), 55, 116, 124, 165n41
LED (local economic development), 117
Louis Berger, 75–6, 81, 168n61

Macedonia, 1, 59, 82, 123, 129, 130, 140, 171n26, 174n9, 177n1
Mihajlov, Andjelka, 121
Milošević, Slobodan, 91, 92–6, 98, 108, 118, 121–2, 125, 126, 148, 161, 169n2
Ministry for Spatial Planning, Construction, and Ecology, 45, 47–9
Ministry of Spatial Planning and Environmental Protection (Sarajevo Canton), 49
MIPD (Multi-Annual Indicative Planning Document), 35, 163n3
MISP (USAID Municipal Infrastructure Support Program), 117
MKT (Ministry for Transport and Communications), 78
Mladić, Ratko, 101, 102, 103, 153
MOFTER (Ministry for Foreign Trade and Economic Relations), 42–5, 47–8, 63, 76
Montenegro, 1, 60, 93, 96, 98–9, 109, 155, 171n26
Montreal Protocol, 106
Mostar Bypass, 66, 70–1, 76–7, 80–2, 85–6, 169n91
multi-level governance, model of, 13–14
municipal level governance
 BiH, 52–5
 Serbia, 110–18

nationalism, 2, 40
NATO (North Atlantic Treaty Organization), 32, 92, 109, 155, 177n1
NEPIs (New Environmental Policy Instruments), 30, 59

NGOs (non-governmental organizations), 5, 8, 54, 73
  BiH, 83–6
  Blagaj and Počitelj, 86–8
  empowerment of, 15, 21, 24
  environmental, 30–1, 35–6, 40–2, 51, 56–60, 61–2, 75
  peripheral participatory role of, 85
  pre-accession funding, 27
  Serbia, 93, 96–7, 106, 111, 113, 141–7
  see also ENGOs (environmental non-governmental organizations)
Nikolić, Tomislav, 94–6, 100
non-state actors, environmental governance, 56–60
Novi Sad, 81, 87, 92, 108, 110, 112, 114–15, 119, 122, 129, 134, 136, 139, 142, 145, 157–8, 172n61
Novi Val, 85, 87
NPAA (National Programme for the Adoption of the Acquis), 105, 107
NPI (National Programme for Integration), Serbia, 105

Obrenovac, 115, 116, 172n53, 172n61
OECD (Organisation for Economic Co-operation and Development), 29
Operation ALTHEA, 34
OSCE (Organization for Security and Co-operation in Europe), 16, 119
Otpor! ('Resistance!'), 93, 148

Pančevo Peace Movement, 122
Patten, Chris, 36
PCDP (Public Consultation Disclosure Plan), 73, 83, 84, 168n69
PHARE (Poland and Hungary Assistance for the Restructuring of the Economy), 97, 163n1
Poland, 30, 163n1
Posavina Canton, 51
PRSP (post-war poverty reduction strategy paper), 68, 166n9
public policy, 11, 19–20, 22, 53

REC (Regional Environmental Center), 51, 55–7, 84, 92, 116, 120
Rehn, Olli, 38, 39
Romania, 15, 18, 30, 112
RS (Republika Srpska)
  Banja Luka, 53, 54, 57, 83–5, 87, 122, 157, 160
  Banja Luka–Gradiška Road, 66–7
  climate change, 44
  EIA (Environmental Impact Assessment) laws, 63–4, 65
  environmental governance, 44–5
  Ministry for Spatial Planning, Construction, and Ecology, 45, 47–9
  Roads Directorate, 66, 73–4, 83
  state and public actors, 73–9

SAA (Stabilisation and Association Agreement), 36–9, 45, 95, 98–101, 154
SAp (Stabilisation and Association process), 1, 2, 8, 10, 34, 37, 147, 154–6
SAPARD (Special Accession Programme for Agriculture and Rural Development), 97, 163n1
Sarajevo Bypass, 66, 67–9, 72, 75–7, 80–2, 84, 86
Sarajevo Canton, 49–51, 77, 164n26
SEA (strategic environmental assessment), 37, 88, 143
SEE (South-east European) states, 13, 41
SEIO (Serbian European Integration Office), 104–5, 169n6–7, 170n11–13, 170n8
self-government, municipalities of Serbia, 110–18
SEPA (Serbian Environmental Protection Agency), 107, 171n26
Serbia, 91–2
  autonomous province of Vojvodina, 108–10
  conditionality, 97–9
  EIAs and environmental governance in, 151–3

ENGOs in, 118–24
EU engagement in, 96–9
Europeanization of, 5–7, 29–31
financial and technical assistance, 97
limits of EU conditionality, 99–103
municipalities, 110–18
NGOs and local communities, 141–7
political future, 158–62
politics and government in post-Milošević, 92–6
state competencies for environmental governance, 104–8
transformation towards long-term partnership, 103–4
Serbian Green Youth, 122
Šešelj, Vojislav, 94–5
SKGO (Standing Committee for Cities and Municipalities), 112–14, 125, 172n50
Slovakia, 15, 30
Slovenia, 92, 148, 160
SNS (Serbian Progressive Party), 95, 96
Socialist Federal Republic of Yugoslavia, 8
Socialist Republic of Bosnia–Herzegovina (SR BiH), 68
Soil Science Institute, Sarajevo, 84
Solana, Javier, 95, 100
Southern Bypass, 70–1
Špirić, Nikola, 38, 45, 163n9
SPS (Socialist Party of Serbia), 94–6, 169n2
SRS (Serbian Radical Party), 94–6, 100
state building, 2–3, 16, 158
state-level capacities
  DEI (Directorate of European Integration), 42–3, 48
  MOFTER (Ministry for Foreign Trade and Economic Relations), 42–5, 47–8, 63, 76
state transformation, 8, 124–5
towards long-term partnership, 103–4
Steinberg, James, 39
Stockholm Convention, 44, 45
Šumatovac, 123, 124, 174n106

TACSO (Technical Assistance for Civil Society Organisations), 35, 36, 62, 97, 163n6, 169n5
Tadić, Boris, 95, 109
TAIEX (Technical Assistance and Information Exchange), 158
Tuzla Canton, 52, 164n30

UDI (unilateral declaration of independence), 94, 101
Una Sana Canton, 52, 81
UNCHS (United Nations Centre for Human Settlements), 92
UN Convention on Biological Diversity, 44
UNDP (United Nations Development Programme), 51, 55, 116, 118
UNEP (United Nations Environment Programme), 92, 116
United Nations Security Council Resolution 1244, 100
University of Sarajevo, 44, 58
USAID, 36, 55, 97, 116, 117, 123–5

Veliko Gradište, 112–13, 172n53, 172n61
Verhagen, Maxime, 100
Vienna Convention, 45, 106
Vojvodina, Serbia, 91, 103, 108–10, 112, 115–16, 122, 126, 134, 139, 145, 148, 171n31, 171n34
Vrnjačka Banja, 112, 116, 172n53, 172n61
Vučić, Aleksandar, 95–6

Washington Agreement (1994), 32, 33
World Bank, 16, 29, 55, 66, 75, 80, 88, 128, 132–3, 136, 137–9, 141, 144–6, 148, 153, 166n9, 175n32, 176n45
WWF (World Wide Fund for Nature), 51, 59, 121

Young Researchers of Serbia, 112, 120, 122, 165n47
Yugoslavia, 1, 4, 32, 56, 75, 91–3, 100, 102, 103, 118, 150

Yugoslavia – *continued*
   conflicts in, 92, 131
   dissolution of, 69, 79, 152
   EU integration of, 154–5
Yugoslav National Army (JNA), 32

Zelena akcija (Green Action), 87, 122, 123, 142
Zelena zona, 123, 124, 174n106
Zeleni putokazi, 123, 124, 174n106
Zenica-Doboj Canton, 51

CPSIA information can be obtained at www.ICGtesting.com
Printed in the USA
LVOW07*1438200815

450911LV00005B/13/P